10/25/2012
• The enlightenment
- technological fixes.
• substances: drugs / caffeine
          as enhancements.

- what do you make of gatica / doping / genetic
                                          enhancements
• communication technologies
   → 3 levels of analysis

Terms:
  coevolution
  agency.

[11/01/2012]

☀ Food distrabution:
      I
      II
      III

Birth contr:
level 1: taking it
2: contr.
3: what happens
   in the long run.

Online education, autonomous cars, jet aircraft, Rail Roads.
Social media.
—Immigration pol., birth control, general contreceptive, Joint stock,
 No school left behind. genetic screening, synthetic biology
 guns, GMO foods. Financial tools. education + technology.
       level 1: tangible technological. shop: cause + effect
                technological fix. simple system

       level 2: system of implamentation

       level 3: ⟨coordinates⟩ the earth effects of the system
          cause → effect. Inpredictiability
          ⟨change⟩

       Food distr:
          1: GMO
          2: additatives
          3: amount of obesity in low income

- what you like about like about this argument

200
Thursday - 75
Friday - 75
Saturday - 75
75
75
75
225

The Techno-Human Condition

The Techno-Human Condition

Braden Allenby
Daniel Sarewitz

The MIT Press
Cambridge, Massachusetts
London, England

For information on quantity discounts, email special_sales@mitpress.mit.edu.

Set in Sabon by the MIT Press. Printed and bound in the United States of America.

Library of Congress Cataloging-in-Publication Data
Allenby, Braden R.
The techno-human condition / Braden R. Allenby and Daniel Sarewitz.
   p. cm.
Includes bibliographical references and index.
ISBN 978-0-262-01569-1 (hardcover : alk. paper) 1. Technology—Social aspects. 2. Human evolution. 3. Biotechnology. I. Sarewitz, Daniel R.
II. Title.
[DNLM: 1. Artificial Intelligence. 2. Attitude to Computers.
3. User-Computer Interface. Q 335]
T14.5.A437 2011
303.48'3—dc22
                                                                    2010039568

10 9 8 7 6 5 4 3 2

To Jonah, Kendra, and Richard, who continue to teach us that technology is neither the answer nor the question, it's just the condition.

Contents

Preface

This book is a product of our good fortune in being selected as Templeton Research Fellows at Arizona State University in 2007 and 2008. Our assignment: explore the implications of radical technological enhancement of human beings—transhumanism—for the environment. We gladly undertook that task, as transhumanism in various guises was of growing interest to us. For example, in May of 2006, we had organized a workshop involving Arizona State University's Consortium for Science and Policy Outcomes and Sandia National Laboratory's Advanced Concepts Group on the public policy implications of emerging technologies for human cognitive enhancement. (The report of the workshop can be found at www.cspo.org.) The workshop touched on the possibility of cognitive de-enhancement technologies as a tool of cultural and even traditional warfare, and on what exactly constituted cognitive enhancement anyway. For example, "off-label" uses of drugs such as Ritalin appeared to most of the people at the workshop to be cognitive enhancement, but caffeine use didn't, and the question of whether Internet search engines were cognitive enhancement or not proved vexing (in part because it required defining "cognitive," and the group—mostly hapless academics who make a living arguing over distinctions between things that look indistinguishable to the real world—could reach no closure even on that).

Thus primed, we approached our charge: How might technology-induced changes in human capabilities affect the environment? This book is the result, but it is nothing like what we—or our sponsors—expected when we started. For one thing, "the environment" quickly proved to be an unhelpful concept for our endeavor, in part because engaging transhumanism requires grappling with the meaning of technological *change*, and mainstream discussions around "the environment" have little that is helpful to say on this subject beyond stale debates between techno-optimistic cornucopians and neo-Luddite catastrophists. But then another weird thing happened: Transhumanism itself turned out to be as conceptually limiting as "the environment," and we soon came to recognize it as, at best, a gesture towards a far more complicated and difficult terrain, where notions of the human, the technological, and the natural seem to become ever more fuzzy and problematic, giving rise instead to . . . what? And that is what this book, drawn from our reflections and public lectures we prepared in the course of our Fellowship, is about. We think.

A note on style: This is an extended essay, rather than an academic treatise, and we have tried to make it readable and enjoyable rather than academic and tedious. For those who want to dig deeper, we offer occasional footnotes and a more complete bibliography at the end of the book.

Our collaboration (in reality a several-year running argument) in writing this book was continually enhanced through our interactions with many colleagues, to whom we respectfully and humbly offer thanks. Among these at Arizona State University are Hava Samuelson, Sander van der Leeuw, Gary Marchant, David Guston, George Poste, Peter French, and Ann Schneider; other perhaps-unwilling co-conspirators in the provocation of our thinking on these matters include Richard Nelson (Columbia), Helen Ingram (UC Irvine), Carl Mitcham (Colorado School of Mines), Roger Pielke (University of Colorado), Steve Rayner (Oxford), Ned Woodhouse (RPI), Mark

Frankel (American Association for the Advancement of Science), and David Rejeski (Wilson Center). Jennifer Brian (ASU) provided invaluable research support to Sarewitz during his Fellowship year; Carolyn Mattick (ASU) did the same for Allenby.

We thank the Metanexus Institute for the grant supporting our Templeton Research Fellowships and the production of this volume, which builds on the Templeton Research Lectures that we presented at Arizona State University in 2007 and 2008. We also thank the ASU Center for the Study of Religion and Conflict, under whose auspices the Fellowship was created and administered, and Carolyn Forbes of the Center for her always-good-natured support of our often complicated and confused path as Fellows. Additionally, we thank the ASU Lincoln Center for Applied Ethics, which integrated a number of their projects with our Fellowship activities to the benefit of all, and Michael Crow, president of ASU, for creating an intellectual environment that permits the likes of this book to emerge.

Finally, we thank our families, who have been subjected to vague ramblings about transhumanism, emerging technologies, human enhancement, Kondratieff waves, and other random phenomena, for all too long. Unfortunately, we can't promise that will stop.

1

What a Long, Transhuman Trip It Has Already Been

Congratulations. You are the proud owner of the latest, new-and-improved-model human brain and body, a version that has only recently become available and that renders all previous models obsolete. Do you think your brain is the same as that of a hunter-gatherer of your species who lived 10,000 years ago? What does it mean that in ancient, oral societies human memory was a principal indicator of intelligence, but we now have search engines that give anyone with a computer access to the world's accumulated memory? Put somewhat differently: Are you as smart as Homer? How do you think you compare to a thirteenth-century peasant, or to Queen Victoria? Queen Victoria could not have even imagined your iPod, and she would have been baffled and probably appalled by what you call music; nor could she have imagined the world's capacity to wipe out smallpox, to control typhus and cholera in European and American cities, or to annihilate itself through an arsenal of 20,000 or so nuclear weapons. To mention just a few of the standard features of your enhanced brain and body, you now come equipped with a fully re-engineered immune system, an up-to-date capacity to distinguish fact from fiction, a completely revised set of cultural assumptions about gender, ethnicity, and sexuality, and, for those of you under thirty, or addicted to i-Phones, a special condensed-language module for instant messaging—all in your own brain and body. Perhaps even more impressive is the amazing range of customized enhancements that some of you have chosen to add to your standard equipment package, including ceramic alloy joints, neurochemical mood modulators, and hormone performance boosters. And if you're cramming for an exam, you may well have just absorbed some psychopharma to enhance your concentration and cognitive function . . . maybe coffee, maybe something more potent and less sanctioned by the U.S. Food and Drug Administration.

You are, in other words, enhanced; some would say *transhuman*, that is, in transition to the next evolutionary phase of humanness. And as such you are also part of a technology-induced evolutionary program that has been going on more or less since the origins of humankind—a program that distinguishes and defines humankind, a program of continuing expansion of the human desire to understand, modify, and control its surroundings, its prospects, and its self, and to couple to the technologies that surround us ever more intimately. From the pre-dawn of civilization, when human tool-making and meat-eating were co-evolving with brain development into the version 1.0 enhanced *Homo sapiens* model almost 200,000 years ago, through the rise of agriculture and the development of early cities with their new capacities for networked human action, through the harnessing of horse power and wind power and water power and the organization of mercantile activities with an intercontinental reach, through the proliferation of the printed word and literacy, and above all through the constant race to develop new ways to exercise military might and kill one's adversaries—in all this business of enhancing the reach and the constitution of our brains and bodies, you are the latest and most advanced iteration.

But perhaps a different game—transhumanism—is now afoot. Until now, some are saying, our application of technology to enhancing our capabilities was largely external: we constructed tools that we could wield to increase our capacity to do things, but as wielders we were essentially fixed in our capabilities. We controlled our external environment, not our internal selves. Even when we did things to enhance our inner capabilities, we did them with external interventions—eyeglasses, education, and the like. Now, we are told, with powerful new genetic technologies on the horizon, with the increasing fusion of human and machine intelligence, and with neuropharmaceuticals, artificial body parts, and stem cell therapies, we are beginning the business of transforming ourselves from the inside out, of exerting explicit and conscious control over our existing selves and our evolving selves in ways that create new opportunities, new challenges, and new ways of thinking about who we are and where we are going. The very notion of what it means to be human seems to be in play. For some people this is a thrilling and wonderful prospect indeed, while others are filled with dread and despair.

But is anything new really going on? Maybe the game is afoot, but what is the game and perhaps more relevant, how can we understand it well enough to play it skillfully, ethically and responsibly? We don't mean these questions to be simply rhetorical: How would you prepare for the Reformation if you were a twelfth-century monk? How would you prepare for the railroad if you were the owner of a general store in Ohio in the 1820s? And if the world we are now making through the technologies of human enhancement really is as complex and unpredictable as we think it may be, what can we do prepare? What *should* we do? And how do you prepare now for a future in which the crucial lessons and values of the past may no longer be sufficient for rational, ethical, and responsible behavior in the future?

As we asked these questions, the 2010 Winter Olympic Games had ended and another Tour de France was about to begin. Amid the determined optimism, the corporate spontaneity, the political scrambling, and the often inspiring athletic competition, eternal questions of doping and fairness remained at center stage. Before the 2008 summer games, *The Economist* had dourly commented "Another Olympics, another doping debate."[1] The Tour de France has become as much a race that pits the latest doping techniques against the newest detection technologies as a contest among the cyclists. Books about doping in baseball are about as common as books about the Iraq War being a mistake. But new themes are sneaking into these debates. One is technological: as gene therapy and genetic engineering replace steroids, bodies are being redesigned rather than merely juiced. Another has to do with the terms of the debate itself, as questions of legality and fairness are giving way to questions of whether genetically engineered athletes are still "real," still "human." If you were born with genes that give you enormous stamina on a bike or on cross-country skis, and I wasn't, why shouldn't I be able to add those genes to myself?

Why not indeed? We have a friend who teaches in law school on questions of law, culture, and emerging technologies. He asks his students how many of them "have close friends or associates" who are taking prescription pills to enhance their cognitive performance.[2] For several years now, more than half of the students have raised their hands—and they have been willing to tell our friend where he can get them.

But if transhumanism is only gene doping and using drugs in ways not approved by the US Food and Drug Administration— off-label uses—why does it suddenly appear as a concept now? Does the concept signal an acceleration of what's been going on anyway—or the beginning of a transformation to something entirely new?

Let us differentiate between two separate dialogs about transhumanism. One involves the ways in which living humans use technologies to change themselves, for example through replacement of worn-out knees and hips, or enhancement of cognitive function through pharmaceuticals These sorts of technological changes are real, although many would argue that such changes have been a part of being human for tens of thousands of years—even if they are now accelerating rapidly. The second dialog positions transhumanism as a cultural construct that considers the relations between humanness and social and technological change. Many people are excitedly talking and writing about the prospects for the technological enhancement of human brains and bodies and a transition to new versions of humanness. The most avid and optimistic of these people call themselves transhumanists. The meaning of "transhumanism" sounds obvious—"between states of humanness"—yet is remarkably difficult to specify. A significant part of the ambiguity arises from one's notions about what it means to be human. This, of course, is contentious cultural territory; after all, without agreement on the meaning of humanness one cannot specify when the technology-enabled leap to transhumanism occurs.

This definitional ambiguity suggests to us that defining "transhumanism" more precisely is less important than understanding the implications of that ambiguity. In other words, "transhumanism" functions more usefully as a lens for observing than as a specimen for studying. If people can't agree on what state we are "transing" from, or to, what then is the deeper issue at stake here?

The World Transhumanist Association originally defined "transhumanism" as follows[3]:

(1) The intellectual and cultural movement that affirms the possibility and desirability of fundamentally improving the human condition through *applied reason* [emphasis added], especially by developing and making widely available technologies to eliminate aging and to greatly enhance human intellectual, physical, and psychological capacities.

(2) The study of the ramifications, promises, and potential dangers of technologies that will enable us to overcome fundamental human limitations, and the related study of the ethical matters involved in developing and using such technologies.

This definition was accompanied by promises:

Humanity will be radically changed by technology in the future. We foresee the feasibility of *redesigning the human condition*, including such parameters as the inevitability of aging, limitations on human and artificial intellects, unchosen psychology, suffering, and our confinement to the planet earth.

More recently, the Association, having rebranded itself as "Humanity+," states on its website (http://humanityplus.org) that its goal is "to support discussion and public awareness of emerging technologies, to defend the right of individuals in free and democratic societies to adopt technologies that expand human capacities, and to anticipate and propose solutions for the potential consequences of emerging technologies," and defines transhumanism in more conceptual terms:

Transhumanism is a loosely defined movement that has developed gradually over the past two decades. It promotes an interdisciplinary approach to understanding and evaluating the opportunities for enhancing the human condition and the human organism opened up by the advancement of technology. Attention is given to both present technologies, like genetic engineering and information technology, and anticipated future ones, such as molecular nanotechnology and artificial intelligence.

The new tone is less insistent, less libertarian, and more sensitive to the need to respond to challenges that emerging transhumanist technologies may raise. The essential focus on the individual and individual capacities remains, however—a focus that we will consider at many points in this book.

Both of the definitions quoted above seem to assume that individual humans are coextensive with technologies that enhance them. But as we suggest below, this assumption carries with it a severe cost, by radically oversimplifying both the challenges that transhumanism claims to address and the institutional and social frameworks within which real people are defined and function.

To start with, the transhumanist assumption that, whatever "human" is, it will only be improved and enhanced—not transcended, rendered obsolete, or even degraded—by the development of transhumanism has the effect of burying both arbitrary values and limits in the definitions of words such as "improve" and "enhance." Many of us may agree, for example, that, with all else equal, enhancing cognitive abilities or reducing pain and suffering is desirable. But as we will consider in later chapters, the technologies that can achieve such benefits may also be potent enough to have other, perhaps less happy effects. Similar questions arise when one contemplates overcoming "fundamental human limitations," for, just as "setting limits" for children may provide the structure that allows them to act more freely and effectively in a social world, so may "limitations" more generally be an important part of what it means to be human, or of how we structure our political and social institutions.

Indeed, despite the reassuring new name "Humanity+" and the apparent effort to move away from dogma, transhumanism remains, in the eyes of many who promote it, a movement. And, as with any political movement, there are significant and growing arguments about what constitutes the movement, and whether it is going in a desirable direction. Some argue in favor of human enhancement on practical, ethical, and even theological grounds; others argue against it as inequitable, futile, or misguided, and even as constituting blasphemy—a primordial sin against the order that God (or Darwin) has established, the Great Chain of Being that gives us all our place.[4]

Transhumanism can also be recognized as just another variety of the technological optimism—one might say hyper-optimism—that has often been conspicuous in Western culture, and especially American culture, having grown out of the Enlightenment commitment to the application of reason to human betterment.[5] Transhumanists, as well as other advocates and visionaries of human enhancement, see many possible avenues of technological development that will continue to drive changes in human capabilities. We will devote little space in this book to consideration of these technological specifics, but they emerge

from the by-now familiar claims of advance in several related
and perhaps converging areas of knowledge and innovation:
nanotechnology, biotechnology, robotics, ICT (information
and communication technology), cognitive science.

—→ The ambitions of transhumanism are comprehensive, extending beyond health and longevity to radically enhanced
intelligence, creativity, and emotional capabilities, conscious
control over the attributes of offspring and the evolution of
the species, and even a greater capacity for mutual understanding through, for example, massively networked brain-
to-brain interfaces. At the limits is total transcendence. As
one employee of the U.S. National Science Foundation writes,
"advances in genetic engineering, information systems, and
robotics will allow archived human beings to live again, even
in transformed bodies suitable for life on other planets and
moons of the solar system."[6] This remarkable statement exemplifies the tendency among transhumanists to extrapolate
from observations about current technology states to breath-
taking visions of immortality, spatial transcendence, and social transformation. Among the better-known examples of this
tendency are the predictions by technical experts such as Hans
Moravec and Ray Kurzweil that, given current accelerating
rates of evolution in information and communication technologies, we will be downloading our consciousness into information networks within decades.[7]

And yet what calls attention to transhumanism is less the
specifics of the agenda and its promiscuous predictions than
the legitimacy that the agenda has garnered. Scientists, engineers, journalists, philosophers, and political theorists, among
others, are discussing the prospects for "redesigning the human
condition." The key claim here is that we are at some sort of
technical threshold where, in the words of a fairly restrained
report titled *Better Humans*, "a new set of possibilities for
[human] enhancement is opening up,"[8] and where these efforts to use technology for human betterment move decisively

inward—into the brain and body and genes—so that, as the journalist Joel Garreau has noted, we become the first species to take control of its own evolution.[9]

But let us first follow the words of the King of Hearts in *Alice's Adventures in Wonderland*, and begin at the beginning. In 2003, the philosopher Andy Clark published a book, titled *Natural Born Cyborgs*, in which he argued that humans have always been cyborgs. In fact, Clark and others claim that our major competitive advantage as a species lies in our brain's unique and innate ability to couple to external social, economic, information, and technological systems in such a way as to evolve distributed cognitive networks. Clark is one of a growing number of scholars arguing not that we will become transhuman, but that we already are transhuman, and have been so almost from the beginning. As our archeologist friend and colleague Sander van der Leeuw has shown, the Paleolithic hunters who over millennia developed increasingly sophisticated sharpened stones for hunting were at each stage in that development themselves cognitively different as well (van der Leeuw 2000). From this perspective, "transhumanism" itself turns out to be a superficial construct that, to us, seems to be of interest principally because it enables continued conflict over the appropriate way to think about being "human," and what the relationship between faith and rational inquiry should be, conducted in familiar frameworks of Western thought.

The recognition that transhumanism may just be what humans do anyway leads, however, to far more interesting questions about the implications of profound technological and social change, and about how poorly we are perceiving, much less adapting to, the challenges posed by those processes in a world already increasingly transformed by human presence. In exploring these questions, we found the swirling arguments over values and transhumanism to have worthwhile illuminating effects. The primary benefit of the discussion, in fact, turns out to be how wonderfully well it illustrates the increasing difficulty of seeing and framing the world we already have created, much less the world that is now coming into being—however intellectually and socially sophisticated we may be. Even as technological, social, economic, organizational, and (yes) cognitive changes coevolve around us, we fall back into classic European Enlightenment terms: liberty, equality, progress, natural order, human "dignity," the Christian Great Chain of Being (and thus the blasphemy of engineering ourselves), and, perhaps above all, the individual as the meaningful unit of cognition, action, and meaning.[10]

Transhumanism is at best a local phenomenon in a far more pervasive reality. All around us is the evidence of our first terraforming adventure—and it is not Mars, but Earth. Indeed, many scientists are beginning to call this era the Anthropocene (meaning, roughly, the Age of Humans). The background to much discussion of transhumanism is a world in which human activity increasingly affects global systems, including the climate and the hydrological, carbon, and nitrogen cycles of the anthropogenic Earth.[11]

And yet we know it not. We are strangers in our own strange land, homeless because we have been turfed out by our very successes. As Stewart Brand put it in his first *Whole Earth Catalog* (1968), "We are as gods and might as well get good at it." So far, we fail that test, and we do so for reasons that the philosopher Martin Heidegger stated succinctly:

So long as we do not, through thinking, experience what is, we can never belong to what will be. . . . The flight into tradition, out of a combination of humility and presumption, can bring about nothing in itself other than self deception and blindness in relation to the historical moment.[12]

We are as gods. This became stunningly clear in 1945, in the New Mexico desert, when a human sun burst into being for the first time. Robert Oppenheimer, standing in the stark shade cast by the flash of the first nuclear bomb, is said to have thought "Now I am become Death, destroyer of worlds." But when

Vishnu, in the Bhagavad Gita, first spoke those words, many centuries earlier, it was as a true god; when Oppenheimer did, he was a mere mortal in awe not of what God or Nature had visited upon us, but what we had built for ourselves—even as that creation equaled the destructive powers that humans had always attributed to their gods. We have since gotten used to, even blasé about, the possibility of nuclear winter, in the way a two-year-old gets used to a loaded .357 magnum lying on the floor within easy reach. We are as gods? No, for we have created the power but not the mind. And as technological evolution continues to outpace the grasp of human intent, we have little time to waste. These are the questions of our time, and they cannot be engaged though flights into tradition.

The more we look at transhumanism as it is currently teed up by proponents and antagonists, the more it reveals itself as something that almost approaches its opposite – a flight into tradition barely disguised by the language of high technology. Rather than some grand prognostication about real future states, transhumanism turns out to be a conflicted vision offering a remarkable opportunity to question the grand frameworks of our time, most especially the Enlightenment, with its focus on the individual, applied reason, and the democratic, rational modernity for which it forms the cultural and intellectual foundation, and the technological New Jerusalem toward which it is flinging us. We accept this opportunity very cautiously. Even if Heidegger is correct and we are increasingly blind to the world we are already engaged in making, it also remains the case that much of modernity is, in our view, desirable—or at least unavoidable. This is our point: As we curl our fingers around the trigger of nuclear weapons, gaze into skies whose dynamics shift inexorably because of our manipulation of the carbon cycle, and unleash technologies that are changing the very essence of our physical and cognitive selves, we are already transhuman. But this is not the kind of transhumanism we thought we were creating, nor is it one we understand.[13]

Like Columbus, we may have started out trying to find the fabled Indies, but instead we have found something new, curious, and unexpected. We ain't (to recall Ellington) what we used to be. But then again, perhaps we never were. And as we go beyond the comfort of old arguments made on behalf of new technologies, we face again the cryptic notation found in some of the old maps of the age of discovery, at the edge of the known: *hic sunt dracones*—here be dragons.

Not to put too fine a point on it, the problem with trying to even figure out how to draw a better map is that people don't understand technology, or the complexity that technology engenders, very well. And this is only going to get worse as humans start redesigning themselves in many ways. So in this book we will proceed in steps. First, we'll try to tease out some crucial understandings of technology by developing a model of its place in the world that can help explain the challenge we, as a species, face, or at least give us a framework for thinking about it. We'll then use our model to explore two pillars of modernity: the idea of the individual and the quest for comprehensibility. And we'll test the model on two major socio-technical systems—railroads and modern military technology—to see how well it works and, equally important, to see if we can use it to think in new and hopefully better ways about the techno-human condition.

We are well aware that the standard approach is to discover deep problems and issues, then offer vague, tentative, or often utterly impractical solutions that pale next to the challenge identified by the analysis—or, instead, simply throw up one's hands in resignation.[14] We hope, in contrast, to end our analysis with some suggestions that have the potential to combine the pragmatic with the radical in confronting the essential dilemmas created by inveterate human ingenuity.

The essence of our response? Stop trying to think our way out of what is too complex to be adequately understood, and

seek the sources of rationality and ethical action in our uncertainty and ignorance about most things, rather than in our
knowledge about and control over just a few things. Add to
that—or derive from that—a degree of psychological and institutional flexibility that acknowledges and dignifies our ignorance and limits. Rehabilitate humility. But first to the essence:
technology.

2

In the Cause-and-Effect Zone

Here at the beginning of the twenty-first century, it is obvious to all that science and technology are continually expanding their reach into the intricacies of human physical and cognitive function. But are we at the brink of something new and different, or are we just pushing further, and perhaps faster, into domains that have already been invaded, and have always been problematic? Certainly the idea that human enhancement is somehow a separate or new or different branch of the larger human technological program is arguable. In what way is a spear, or a bicycle, or a book, or a telephone, or eyeglasses not a human enhancement? How are such things different from the sorts of things the transhumanists and their more measured allies have in mind? Much is made, for example, of the alleged potential of emerging cognitive and genetic technologies to embed enhancements both in our brains (thus internalizing them in the organism) and in our genes (thus propagating them to our progeny). But the fact that, say, our great-grandfathers owned and rode bicycles, as we do, tells you that temporal propagation of enhancement can perfectly well be embodied in the external technologies, not just through our genes, while the apparent fact that you can never forget how to ride a bicycle, or how to read, tells us that allegedly external technologies do in fact have an enhancing effect on our internal capabilities.

It turns out that we can't even have the right conversation if we don't first get past the Cartesian dichotomy between mind and body, which we children of the Enlightenment have all absorbed without question. Sure, there are some questions that can be addressed best by thinking about the individual (and the individual's brain) as separate not just physically but metaphysically from external reality (for instance, questions about the genetic bases of bipolar disorder or schizophrenia, although even there external triggers are an integral part of expression of genetic potential). But when it comes to understanding the human—and questioning the transhuman—the Cartesian duality can be downright misleading. If you were taught a fact in 1990, you memorized it; if you need a fact now, you google it. (Turning a corporate brand into a verb is one flag of a socially interesting phenomenon, and in this case "to google" is a profound statement about important, and very new, changes in cognitive systems.) The Internet becomes an increasingly important part of one's overall memory strategy. Sure, you can still remember facts, but why tie up your limited personal cognitive capacity with an unnecessary function? You just make yourself less competitive with the folks that don't. This is not conceptually new, of course; before the Internet, books did the same sort of thing: I didn't have to memorize Shakespeare's plays; I merely had to remember where I had put the book. The history of our species is a history of redesigning ourselves, of fuzzing the boundaries of our inner and outer worlds.

Most experts on early human evolution agree that primitive tools and human brains co-evolved; that the imaginative capacity of the tool-maker was both a product of and a requirement for the development of more effective stone tools and more rapid innovation. Education is, of course, a conscious process of brain modification, and culture is a process of passing such modifications from generation to generation. Edward Jenner started modifying immune systems with cowpox pus in 1796, and the more dangerous practice of variolating with smallpox pus had apparently been used in China since about 1000 A.D. The rise of the printing press and the widespread distribution of printed vernacular texts created an unprecedented and far-flung cognitive and information network, one that had profound culturally transforming impacts (later amplified by the rise of telegraph and telephone networks). In the nineteenth century, German doctors were outfitting war amputees with prosthetic arms designed to fit directly into mechanisms used in factories to control machines, thereby blurring the human/machine boundary. Thus, it isn't clear to us that we now are crossing some domain that humans have never entered before, a domain that demands a new kind of debate or raises new moral considerations and dilemmas. Nor are claims of miraculous advances to come unfamiliar, or the counter-claims of impending troubles, or the moral argumentation for and against. It has been the nature of scientific and technological advance to provoke ardent support and committed opposition on grounds ranging from the spiritual to the pecuniary; and the transformational power of technology has shaken societies to their roots at many points in human history, and will surely do so again.

But transhumanism and the more general goal of human technological enhancement aren't familiar merely because of their connections to an ongoing process of technological transformation of society that seems very much interwoven with the human condition itself. In particular, one need not look too deeply into the language used to promote transhumanism and human enhancement to recognize an agenda for human betterment that in other contexts marks the domain of faith and spiritual practice. Transhumanists explicitly embrace the pursuit of immortality, of human perfectibility, of dominion over nature, and of transcendence over the limits that time and space impose on the individual. Transhumanism also shares with many religions a millenarian, apocalyptic vision of the future day when paradise will be gained or regained, although for transhumanists that day will come when humans (or at least human minds downloaded into computers), either out of necessity or by choice, will leave Earth and expand their domain and dominion in the solar system and toward the infinite, their worldly spirits still, apparently, intact. This is exactly the future seen by Moravec, Kurzweil, and others.

Immortality, perfectibility, dominion, transcendence—this mimicking of religious goals by technological visionaries is no coincidence. In his book *The Religion of Technology*, David Noble (1998, p. 52) details how science and technology were viewed by seventeenth-century English scientists (exemplified by Francis Bacon) as the tools for regaining paradise—for "fulfillment of the millenarian promise of restored perfection." Scientific knowledge will allow humans to "stretch the deplorable narrow limits of man's dominion over the universe to their promised bounds" and attain a "true vision of the footsteps of the Creator imprinted on his creatures."[1] Sounding themes central to transhumanism, Bacon in 1627 foresees, in his final, utopian work, *New Atlantis*, "the prolongation of life: the restitution of youth . . . the curing of diseases counted incurable . . . transformation of bodies into other bodies . . . making of new species . . . force of the imagination upon . . . another body."[2]

The similarities between Enlightenment enthusiasms and some of the proclamations made on behalf of technological enhancement of humans are positively spooky. We will, the technologist Ray Kurzweil writes, "transcend [the] limitations of our biological bodies and brains. We will gain power over our fates. Our mortality will be in our own hands. We will be able to live as long as we want. . . . Our technology will match and then vastly exceed the refinement and suppleness of what we regard as the best of human traits."[3] The claim here is not just one of material betterment, but of improved humanness (a claim that ought to set off alarm bells, given the history of the eugenics movement in the early twentieth century). Machine interfaces, neuropharmaceuticals, and genetic modifications can all help. For example, the biophysicist Gregory Stock writes: "The arrival of safe, reliable germline technology will signal the beginning of human self-design. We do not know where this development will ultimately take us, but it will transform the evolutionary process by drawing reproduction into a highly selective social process that is far more rapid and effective at spreading successful genes than traditional sexual competition and mate selection."[4] We are, it turns out, in neither God's nor Darwin's hands, but in our own. In an extraordinarily strong statement of faith in the power of reductionist science, the philosopher Nick Bostrom explains how this works: "The difference between the best times in life and the worst times is ultimately a difference in the way our atoms are arranged. In principle, that's amenable to technological innovation. This simple point is very important, because it shows that there is no fundamental impossibility in enabling all of us to attain the good modes of being."[5] Here is the technological enhancement of humans in direct competition with religion over the dispensation of our character as individuals and as a species.

Now, if you spend some time, as we have done, reading about transhumanism and human enhancements—the debates about what is technically plausible and what is not; about what is morally acceptable and what is not; about who will benefit and who will be left behind—you will find that pretty much every feasible position has been staked out and defended vociferously and sometimes even capably. You can read, for example, the bioethicist John Harris's *Enhancing Evolution* and the political philosopher Michael Sandel's *The Case Against Perfection*, and find serious, carefully constructed arguments that lead in antipodal directions. Similarly, there are some, including the theologian Philip Heffner, who find a sympathetic relation between the aims of technological enhancement and religious practice, and others, among them the philosopher Alfred Borgman, who see fundamental and corrosive conflict. You can also find diametrically opposed arguments about whether technological enhancement is or is not an obligation of democratic societies, whether it will improve democracy or undermine it, and whether it will improve justice and equality or erode them. Yet the various combatants do seem to share what seems a rather incredible assumption—an assumption that makes their disputes possible in the first place, and thus legitimizes the debate and all the attention it attracts. Everyone seems to accept that something new is happening, centered around emerging prospects for changing humanness, for steering its future, through the achievement of new levels of direct control over the physical and cognitive performance of human beings, including the controlled biological evolution of performance standards, the direct intervention in brain function, and the gradual hybridization of human and machine intelligence. The starting point for these diverse moral and philosophical treatments is that emerging technological potentials make humanness—however one wants to define it—an appropriate subject for intentional design in a way that is unprecedented.

In our own overly literal way, we want to begin our exploration by testing this assumption and its implications for the meaning of technological enhancement of humans. When it comes to improving humans, to making us better in any meaningful sense of the word, are there good reasons to think that new technologies can do the job—or, at least, can do it better than, say, religion or politics? In other words, we want to consider not what is technically correct or morally right about technological enhancement itself—which only leads to a hotly contested but perhaps ultimately unproductive boxing arena as people defend their respective normative corners—but what is, and what is not, operationally feasible in this program of making humans better. The type of "better" we mean here is not about cured diseases or healthier lives, but about, as the transhumanists promise, "redesigning the human condition"— making humans better than they are across all dimensions of the world made, inhabited, and experienced by humans. John

Harris (2007, p. 2) writes: "Enhancements of course are good if and only if [they] do good, and make us better, not perhaps simply by curing or ameliorating our ills, but because they make us better people."

Those who have staked out positions of opposition to, or discomfort with, the technological program for enhancement of human capabilities have mobilized several types of arguments. First there are those who call upon some fundamental sense of appropriateness, of received human dignity, of what is natural, right, and sufficient in our world, to question the wisdom of the transhumanist agenda. Leon Kass, former chairman of the President's Council on Bioethics, is perhaps the leader of this approach on the right side of the political spectrum, and the writer Bill McKibben has made a related case from the left. Another line of critique, led by the political philosophers Francis Fukuyama and Michael Sandel (respectively right and left of center), suggests that changing fundamental aspects of the constitution of humans will threaten the fabric of our social and political institutions in ways that are likely to be negative. Other arguments focus on questions of distributive justice, risk, and cultural erosion.

In contrast with this diversity of critiques, the starting place for most of the arguments that favor a libertarian approach to technological enhancement of humans is a strong defense of the rights of individuals to make decisions about their own capabilities and those of their children or children-to-be. This resonates well in modern market democracies, in which individual autonomy is a fundamental value. And we have to acknowledge that, having subjected ourselves to a lot of this debate, there is a reductionist rigor to the individual-rights argument that is simply not available to those whose reservations about technological enhancement are based either on some sense of fundamental human authenticity or on forebodings about future hypothetical consequences. Concrete statements about individual rights are pretty easy to make, test, and defend;

concepts like authenticity and dignity are much more slippery, and lend themselves to sweeping normative pronouncements without the beneficial constraint of having to consider real effects on real people.

The individual-rights defense also allows the advocates of aggressive human enhancement to distinguish their ambitions from repugnant past efforts to engineer human improvement through coercive, state-sponsored means—especially the eugenics movement in the United States and elsewhere in the early part of the twentieth century and the genocidal eugenic ambitions of Nazi Germany shortly afterward. If the means of human enhancement are applied at the discretion of the individual and administered through the democratically regulated economic market, then enhancement becomes an expression of freedom, not repression, and a path to diversity, not homogeneity. Repression thus becomes the tool of those who would like to prevent individuals from choosing to enhance themselves (and indeed there are some who call for state bans on various technologies and relevant areas of research), and the transhumanists become the guardians of individual freedom.

But if the goal of human enhancement is better humans, and better humanity, then the individual-rights perspective faces a serious scale-up problem. For one thing, people are not simple summations of their individual traits; knowing that a person has an enhanced trait, perhaps an implanted memory chip, doesn't really tell us anything useful about who that person is. Second, humanity—the aggregation of humans—is not a simple summation of a bunch of humans, much less of a bunch of human traits. As compelling as the individual-rights case for pursuing human enhancement might seem to be, the human enhancement program cannot be about individuals alone, because the enhancements of traits and abilities are benefits only insofar as they allow us to act more effectively in a world of other people, where social, cultural, and institutional structures help to determine what counts as effective. In most cases, arguments based on the individual are fatally decontextualized, in the following way: Although strong reasons in favor of allowing individuals to enhance their physical and mental attributes (and those of their children and their prospective children) can be rooted in an individual-rights perspective, whether these enhancements actually add up to an improved life, an enhanced life, or a better life—whether they even improve the chances of attaining such—is only weakly coupled to the enhanced attributes of the individual. To say that we have enhanced—made better—the particular trait of a particular human has no necessary predictive power in terms of what we have made better at the level of the whole person, and of humanness more generally.

This position must be distinguished from the better-known and more sophisticated argument that Francis Fukuyama and others make about the dangers of messing with human nature. Fukuyama describes an ineffable quality of humans that is not reducible to any particular function or trait—he calls it "Factor X" to indicate that there is something special that we can recognize as an essence of humanness, and, though this essence isn't something we can put our finger on, it is something we want to nurture and protect, particularly against transhumanist technologies. This approach, of course, raises questions of power and choice. Who gets to define what the essence of humanness is that must be protected? To what extent can particular belief systems coercively impose their view of humanness on others? But even if one is sympathetic to the perspective that certain technologies threaten some shared sense of essential humanness, the holistic mushiness of this view cannot easily compete with the more philosophically reductionist arguments based on the rights of individual choice.

So here we are in the maw of an unhelpful, even incoherent discussion pitting the integrity of a vague and inescapably normative shared human essence against the concreteness of an individual human right. But what if we ask a different sort of question? What if we forget about trying to defend, or even posit, some particular notion of "human nature," and instead look at the core claim made by advocates of transhumanism and human enhancement: that technology, through the choices individuals make about their own enhancement, will make humans, and humanness, better.

The argument that individual choice and personal freedom are the essence of transhumanism implies a testable prediction: that the more enhanced among us will be the more free, the more capable of exercising choice. So let's start the testing: Who are the most enhanced individuals in the world today—enhanced both physically and cognitively with the latest technological advances? Almost certainly they are America's soldiers in Iraq and Afghanistan, with their smart weapons, their body armor, their night-vision goggles, their special diets, their training in and integration into remote robotic combat systems, and, we would suspect, their ingestion of neuropharmaceuticals such as modafinil to keep them alert even when deprived of sleep for 36 hours. Who among us is more enhanced? Yet who among the less enhanced of us reading these words would choose to trade places with them? We don't mean this at all glibly. We would prefer that they weren't there at all. But since they are there, we want them (as do most Americans) to have every enhancement possible to improve their chances of a safe return home. In any case, the fact that soldiers in combat are the first individuals to receive the benefits of many emerging types of physical and cognitive enhancement shows that the enhancement program is not just about individuals choosing to improve on their humanness, it is also about institutional and political settings in which human enhancements are being used to advance goals that have nothing to do with individual expressions of liberty in the pursuit of life goals. Indeed, these are likely to be the principal settings of the human enhancement program.[6]

Enhancement at the individual level need not lead to an enhanced individual or to an enhanced society. Consider a drug or a brain implant that improves the ability to concentrate, as the prescription drug Ritalin does. Now, individuals may take this drug—indeed, they are taking it—to, for example, improve their performance on university exams—pretty much everyone wants to perform better on exams. But high performance on exams is just one attribute of a person who might otherwise be a jerk. The point isn't that jerks shouldn't be allowed to perform well on exams, it is that making any statement one way or the other about the value of better concentration is hard to do if one is considering people as entities, rather than as aggregates of individual, enhanceable traits. By what definition is a jerk with better concentration a better person than he was before? If lots of jerks improved their concentration, the cumulative effect on the rest of us might well be unpleasant. There is an aggregation problem at the individual level. Enhancing individual traits or capacities is a piecemeal project that tells us nothing at all about what society can expect from the individual who is being enhanced—or about what society can expect from millions of such enhanced humans.

And what happens when lots of people start improving their concentration? Well, for one thing, presumably lots of people can benefit from whatever pleasures improved concentration might deliver (although perhaps they would also be deprived of other pleasures, such as daydreaming their way through long, boring lectures[7]). But to the extent that people want to improve their concentration to make themselves more competitive at, say, exams, or in the courtroom, or on the tennis court, the benefits of improved concentration across the whole population are diminished, as when everyone in a crowd stands on tiptoe to get a better view. So there will be incentives to hoard the benefits, and there will be enhanced expectations of better performance among individuals—expectations that will then be vitiated for many or even most individuals because others are doing the same thing, thus perhaps creating more disappointment than without the enhancement. We see this phenomenon at work in American society today; for example, competition for admission to elite universities (and even elite New York City kindergartens) has become unpleasantly intense as parents do everything in their power to improve their children's competitive positions. Some might still interpret this sort of phenomenon as making society better by improving overall performance, even while perhaps stimulating new heights of individual neuroses. In any case, it turns out that the direct avenue of technological effect on an individual trait—the enhancement of individual concentration—feeds back into the individual through aggregate consequences that can undermine the original intent behind the particular enhancement. This phenomenon, sometimes called "social limits to growth,"[8] is neither surprising nor uncommon; it's exactly what we ought to expect when people are enhancing themselves in order to gain a performance advantage that others will seek too. And, as we shall see, it offers an interesting way to reconceptualize the relationship between technology and society.

Are baseball and bicycle racing better because individuals are enhancing their physical abilities and performing better? What is the measure of "better" here? Did spectators and athletes enjoy themselves less in the past, when competitors were less enhanced? At this point, the judgment of society is that certain types of enhancement, such as the use of growth hormones and steroids, diminish the value of competition. The individual-rights-based response is that our expectations of what is normal or acceptable—for example, the rules of a particular sport—are arbitrary and always in flux. The vitamins, exercises, and nutritional regimes that athletes benefit from today without public protest make them radically enhanced relative to athletes of the past. How, then, can we justify opposing the next level of enhancement? But such arguments miss the deeper question: If all the enhancements do is keep raising the level of performance, and our expectations along with it, then from what Sisyphean frame of reference can we say that things are getting better? "Better" begins to seem a lot like a synonym for "more"—and, indeed, as in the case of education, one might suspect that the real forces behind enhancement are efficiency, productivity, and growth, not the higher values that are so readily bandied about. After all, our students are not enhancing themselves in a vacuum; they are competing to be hired by firms that value their increased productivity and economic output, not their "happiness" or "freedom," in a society that measures its achievements in terms of gross domestic product and comparative advantage over other societies. The United States and China are not investing in enhancement technologies out of altruism, but because the technologies are supposed to offer comparative advantage in the perpetual jockeying for cultural, economic, and geopolitical dominance. We are not trying to argue here that efficiency, productivity, and cultural authority are in some way "wrong"—indeed, that might seem rather rich coming from two American authors. But we are questioning the coherence of an individualist perspective on transhumanism.

The test of the real world, then, indicts both poles of the transhumanist debate. To the individualists, it says that those who are most enhanced are being enhanced by large and powerful institutions, or coerced by large and powerful economic and cultural processes, not as a result of free individual choice—and who among us is not subject to such pressures? Indeed, academic professionals are cognitively enhancing themselves in large and increasing numbers; we, your humble authors, may not be far from the day when, if we want to remain competitive, we will have to enhance ourselves. For the record, if it is not clear from the text, we are not yet doing so, unless one counts caffeine.[9]

But neither is it possible to accept the alternative claim that individuals will or should be disgusted with transhumanist technological options, and not enhance. One cannot help but note the high demand, among those with high disposable incomes, for plastic surgery, botox treatments, and the like—enhancement technologies that, at some risk, offer better looks but not increased life span, better health, or even, necessarily, enhanced quality of life. And those enhancements are merely cosmetic. The doping arms race between high-performance athletes seeking a competitive edge demonstrates the zeal with which enhancement options are adopted when they offer the potential for improved performance (and glory, and endorsements). Some people may think enhancement is against "human nature," or immoral, or a violation of religious or natural law, but if real enhancement is within reach, even if legally or medically risky, the data all seem to indicate high demand. If enhancement is to be stopped, it will have to be stopped by society's actively preventing people from doing what they seem to want to do. The argument against enhancement on these grounds, therefore, does not seem to be an appeal to the public, which appears fairly immune to such sentiments, but to the State, in that it asks for and justifies an authoritarian mandate for reasons that are essentially ideological, even theological.

Moreover, in a world dominated by large and competing institutions—multinational corporations in constant combat for market share, nation-states worried about security, cultures jostling for global dominance—we can make two predictions with considerable confidence. First, the beneficiaries of enhancement will generally not be individuals, but institutions; as a corollary, the drivers for enhancements will be economic efficiency and competition for military and cultural dominance, not quality of life or "better humanness," even if we knew (or could agree on) what the latter was. Second, particular enhancements cannot be viewed in isolation; they are changes in highly complex and adaptive systems, and the relatively primitive utilitarian tests of the transhumanists, and the highly normative moral or ethical tests of the anti-transhumanists, are each grossly outmatched by the reality at hand. Modern notions of the individual, of progress, of freedom, of rational choice, and of ascertainable links between causes and effects undergird a discussion of transhumanism that could generously be seen as besides the point. Yet there are complex phenomena and deep issues in play here, even if the current debate does not explicate them, and it is in search of these that we now go.

3

Level I and Level II Technology:
Effectiveness, Progress, and Complexity

Could there be any better mirror into a person's soul than the person's views on progress? Which litany best supports *your* worldview: the eradication of smallpox; the lifting out of poverty of hundreds of millions of people in South and East Asia; the economic and political integration of dozens of European nations that for centuries were at one another's throats; the defeat of Nazism, Stalinism, and Maoism; the creation of an amazingly egalitarian global information network via the Internet? Or are you more comfortable with The Bomb, AIDS, climate change, continuous concentration of global wealth, a billion malnourished people with no access to clean water, and information overload? What a great subject for academics to argue about from now until eternity. From the inevitability of progress to its impossibility; from its invention as a modern ideal to its persistence throughout history; from its embodiment in scientific truth-seeking and technological advance to its social construction as nothing more than a contextual illusion that justifies particular ways of being and acting, progress can shoulder just about any philosophical, cultural, ideological, or statistical burden that we want to place upon it.[1]

"Progress" is central to our interrogation of transhumanism for two reasons. First, it is an important Enlightenment motif: progress is possible, and applied rationality is the way to continue progressing. Second, at the core of discussions on progress we often find technology, and for good reasons. There is a kind of irreversibility to technological change that makes it a particularly inviting frame of reference for thinking about what progress might actually mean. Technology offers a continually, if unevenly, expanding domain of increasing human control and power in the world, and in the process technology continually transforms the natural and social worlds. Technology instantiates the natural laws that science discovers, and thus represents the worldly application of the scientific search for truth (even if, in many cases, technological application precedes and enables the scientific discovery of the laws). Technology also substantiates the imagination of humanity in our inventions, and introduces novelty into the world—capabilities and artifacts that never could have existed before. For many, technology embodies the modern ideal of applying rationality to the betterment of humankind.[2]

Modern society long ago lost its innocence about technology and progress, of course. We have gone from technology as a particular artifact or machine that just does its job to understanding that it emerges from social systems and thus necessarily reflects, internalizes, and often changes power relations and cultural assumptions. We recognize that social systems are in reality techno-social systems, that these systems impose certain orders of behavior on our lives about which we have little choice, and that these systems lock in paths of dependency that make a mockery of human agency—just try decarbonizing the global energy system! Techno-social systems also make possible hierarchies of expertise, influence, and exploitation—who, today, can argue with an auto mechanic? We know that technological systems are now as complex, pervasive, and incomprehensible as natural systems; in fact we know that the distinction between technological and natural systems is no longer very meaningful. We know that the dependence of modern market economies on continual growth means that we have to keep inventing and consuming new technologies, whether we really need them or not—indeed, it is not clear what "need" means in our modern framework, Abraham Maslow to the contrary.[3] Moreover, this process of continual innovation, productivity enhancement, and economic growth leads to apparently unavoidable spasms of severe unemployment and socioeconomic disruption and transformation, along with the wealth creation that seems to have become an underpinning for civil stability.

Technology, that is, seems at once to render the idea of progress more concrete and more perplexing—and more suitable for academic research and debate. Writing in the early 1930s, Lewis Mumford called this the "ambivalence" of the machine, which, he observed, "is both an instrument of liberation and one of repression." "It has," Mumford continued, "economized human energy and it has misdirected it. It has created a wide framework of order and it has produced muddle and chaos. It has nobly served human purposes and it has distorted and denied them."[4]

In this light, then, the area of innovation lumped under the term "human enhancement" can be understood as simply the latest version of the technology-and-progress Rorschach test, and transhumanism as a claim for old-fashioned, technology-induced progress—for things, generally, getting better because of the development and use of technologies, in this case applied directly to making human bodies, genomes, and brains better than they have ever been before. And perhaps because we ourselves are the central objects of innovation, transhumanism provokes engagement with all the ambiguity and ambivalence of technological progress—in spades. We are redesigning our selves, or so it is claimed, with ardor on one side and with grave reservations on the other. What is going on is nothing novel at all; humans have always been in the game of transforming themselves with technology. What is going on is radically new; we are on the threshold of intervening in our own evolution, in the exponential expansion of our own cognitive capacities, in an irreversible blurring of the human/machine boundary. The

benefits to humans will be wonderful as we achieve new levels of intelligence, creativity, agility, even perhaps wisdom. The
threats to humanity are profound as we alter our natures in
ways that may erode the foundations of society, challenging
our commitments to justice, democracy, and the very notion of
human dignity. And so on.

Amid all this ambiguity, ambivalence, hope, fear, and philosophical mud wrestling, can we say anything at all about what
really is likely to happen? Perhaps, but we will have to reject
the idea that we'll be able to make specific predictions about
the technologies themselves, and we'll have to reject the substitution of doctrine for analysis. Instead, let's begin with the
obvious point that, if this human enhancement thing goes anywhere, it will be because people use the technologies. And let's
enter the fray by positing the existence of something that no
person of any intellectual sophistication could possibly believe
in anymore: technology itself. For example, we want to claim
that the watches we all wear embody a technological function
that we find useful: letting us know what time it is. In acknowl-
edging this function we're not at all suggesting that, say, the development of the chronometer in the eighteenth century didn't
advance the pace of colonialism by allowing more precise marine navigation, or that we weren't manipulated into buying a
particular brand of watch by clever advertising, or even that
our need to know the time isn't a cultural expectation that
assaults individual autonomy but is, rather, a product of industrial capitalism rooted in the development of the railroad.
(On that, see the next chapter.) All these things probably are
true. Even so, watches help us navigate our days because they
reliably do what they were designed to do and what we want
them to do.

Another important approach we want to take that differs
from many discussions about enhancement and transhumanism
is to focus on things that are happening now in the real world,
so that the opportunities and dilemmas are not distanced from

our sense of the difficulties of the world we live in but instead are conditioned by a recognizable context. We also want to avoid distracting debates about what is or is not technologically possible in the future. In a world already suffused with technology, people will be basing their choices about new technologies in part on their experience and expectations of what, in their experience, works and what doesn't, rather than on the unbounded promises made about technologies that don't yet exist. These choices then help determine how the technologies will evolve, in ways we cannot possibly know far in advance—especially since it isn't entirely clear how our individual choices, summed up across these complicated social, cultural, and economic systems in complex ways, affect the evolution of technology to begin with.

Yet, just as we want to ground our discussion in things happening now, it seems reasonable to expect that, as future dilemmas about human enhancement issues unfold, they are going to feel much like the current ones: you aren't going to wake up one morning and suddenly find yourself in a world where you can buy computer-brain interfaces (known as CBI to the cognoscenti) that will boost your IQ 100 points,[5] or genetic modifications that will render you impervious to aging. To the extent that such technologies—staples of the transhumanist debate—are even remotely on the horizon, they will be approached slowly and unevenly, with front-page claims of amazing advances one day and page-seven revelations of disappointed expectations a year later.[6] Pilot tests fail, entrepreneurs are over-optimistic, new discoveries in completely unrelated fields result in sudden shifts in what seems possible and economically feasible, and complexities and difficulties arise that can derail the seemingly simple and predictable. Moreover, technology is not just a matter of innovation; it is also a matter of adoption by a critical mass of users, and it co-evolves with cultural, economic, political, and other domains, each of which continually affects, and is in turn affected by, the others.[7]

With this in mind, let us begin on what we hope is firm ground: Making and using and improving on technologies is—like eating, sleeping, lying, procreating, and imagining—something that people do; it is part of being a person. "The age of invention," Mumford wrote, "is only another name for the age of man."[8] And one aspect of this innateness is that, at any given moment, humans are utterly and irrevocably dependent for their survival on some set of technologies and technological functions. Of course the nature of that dependency changes over time, sometimes in radical fits and starts, but it remains, nonetheless, a condition of our condition. And this dependency means that assessments of future technologies are always relative to the current technological state, not relative to some pre-technological or non-technological state. Technology is always part of the context, and to see it as exogenous to the human in some über-Cartesian spasm of solipsistic individualism is simply wrong.

We want to reinforce the point about functionality. Technologies are meant to do things, to accomplish particular goals or tasks. Technologies are, in effect, cause-and-effect machines, linking a human intent to a particular consequence via the embedded function of the technology, and often doing so with very high reliability—often (though not always) much more reliably than could be achieved with an older technology aimed at the same function, or without the technology at all. And that a technology emerges from cultural and political choices, or that many technologies seem to serve no useful purpose other than to advance the profits of the producer, or are designed to do things like kill people, or end up doing things that they were not originally designed to do (like using your cell phone to tell time), is not problematic for this assertion of bounded functionality.

Our purpose here is to create a simple taxonomy of levels of technological function to allow a little clearer understanding, and a little less confusion, about the differences between toasters and nuclear weapons. Let us begin this process in a somewhat academically incorrect manner by connecting technology to a very modest notion of rationality: We suggest that when people make a decision, they usually intend to achieve the objective at which their decision is aimed. At the level of individual use, we claim, technologies are volition enhancers: they often give us a better chance of accomplishing what we want to do than would otherwise be the case. To take a trivial example, suppose that Allenby or Sarewitz, as often happens, is asked to present a lecture in a far distant locale. He could certainly decide to walk or bike to the lecture venue. But even if he were carefully to plan such a cross-country trek, he would be much less likely to arrive in time than if he instead chose (as he usually does) to avail himself of one of our infuriatingly unreliable airlines. By condensing as much of the cause-and-effect involved in a cross-country trip into a single, contained technology that is, in fact, extraordinarily reliable—the jet aircraft itself—one maximizes the potential for achieving the intent of the decision to be where one wishes, when one wishes.

But this familiar example raises another point: although particular technologies are reliable cause-and-effect machines for our use, they are also components of complex systems whose connections to the functionality of the individual technology may not be obvious. Technologies, that is, inhabit two rather independent realities. First there is the reality of the immediate effectiveness of the technology itself as it is used by those trying to accomplish something—for example, the jet airplane that carries one with incredible reliability from A to B. We call this Level I. The other reality is that of systemic complexity—for example, the air transportation system, which seems in many ways to be the embodiment of irrationality and dysfunction, with its insane pricing system, the absurd inefficiency of its boarding and security processes, the continual delays, and the increasing impossibility of spending one's frequent-flier miles, not to mention the almost continual financial insolvency of most airlines. Here we have a hugely sophisticated yet physically discrete, tangible, and recognizable technology that very effectively meets our requirements (that's Level I), but it is embedded in a larger system, the air transport network, that is itself a complex socio-technological system, infinitely less predictable and more complicated than the jet airplane itself. We'll call this Level II technology.

Level II is less bounded than Level I. It includes subsystems—airline corporations, the government security apparatus as applied to air travel, and market capitalism in route pricing, to name a few—that, acting together, create emergent behaviors that cannot be predicted from the behavior of individual, Level I aircraft units. For example, at Level II one gets such phenomena as technology "lock-in," which occurs when economic, cultural, and coupled technology systems coalesce around a particular way of doing something—as we see in the automotive industry, where hydrogen fuel cell propulsion technology is feasible today, but the energy-supply infrastructure necessary to support it is not. The gasoline internal-combustion engine is thus "locked in" by the economic interests of the suppliers of petroleum fuels, the physical infrastructure of pipelines and gas stations, the interdependency of gasoline internal-combustion engines and gasoline, and the cultural role of fossil-fuel-consuming automobiles. Computer chips and software technologies are similarly interdependent, tending to lock one another in, and to advance in lockstep as changes in each make possible corresponding developments in the other. Lock-in does not imply that technological change is impossible, of course—merely that it strongly tends to follow paths that reflect past system states.

In short, we are now back in the comfort zone of ambivalence, ambiguity, and mud wrestling. This Level II system complexity that accompanies a reliable Level I technology raises another familiar challenge: the likelihood of unintended consequences. Technologies do not act in isolation; they are connected to other technologies, and to social and cultural patterns, institutions, activities, and phenomena that may interact in ways that no one is able to predict or control. So the functionality embodied in the airplane that reliably carries one from coast to coast is also implicated, for example, in the rapid spread of exotic infectious diseases such as AIDS and SARS, and in terrorist attacks that marked a major inflection point in recent U.S. and even world history. Technologies often surprise us because they introduce into society novel capabilities and functionalities whose uses are constantly being expanded and discovered—capabilities and functionalities that interact with other technologies, and with natural and social phenomena, in ways that cannot be known in advance.

But even here the short-term complexities of air transport systems and networks are more bounded than the longer-term implications of the technology system taken as a whole. For example, the automobile itself is a Level I technological solution to the problem of getting from one place to another effectively, relatively safely, at a time and rate of one's choosing. The automobile as a Level II technology network is different; it creates malls, suburbs, highway systems, and petroleum-delivery infrastructures. Moreover, cars in networks create emergent behaviors that, at least temporarily, subvert the usefulness to the individual of the car as artifact. (We call one such emergent behavior a traffic jam.) But as the basis of a technology constellation that fueled a stage of economic evolution in the West, the automobile did far more: it co-evolved with significant changes in environmental and resource systems; with mass-market consumer capitalism; with individual credit; with behavioral and aesthetic subcultures and stereotypes; with oil spills; with opportunities for, and a sense of, extraordinary human freedom, especially for women who were otherwise trapped in their houses by social and economic patterns.[9] We'll call this Level III.

The divisions between these levels of technology are not necessarily clear, both because any such schema is necessarily somewhat arbitrary and because the level at which a technology manifests itself depends in part on the goals we attribute to it—one artifact can, depending on what goals, policies, and social systems we are interested in, be a part of a Level I, a Level II, or a Level III system. If one thinks of a vaccine as a means of reducing levels of infection, it looks like a Level I technology; if one thinks of it as a means of improving economic growth, it looks like a Level II technology; if one thinks of it as a part of long-term demographic trends and subsequent political and social evolution in a developing country, it looks like a Level III technology. (For example, do vaccines contribute to a "demographic bulge" that feeds a pool of unemployed and disaffected teens who can be radicalized by terrorist organizations?) Same artifact, different system boundaries implied by the analysis. Moreover, there is a major perceptual difference between the first two levels and the third: A driver is familiar with her car, grumps at the congestion on the roads on which she must travel, is aghast at the cost of fuel even while expecting it to be widely and safely available, and stops at a drive-through fast food joint for coffee and doughnuts on her way to work. She is far less likely to exclaim at the wonder of personal credit, "big-box" retailers selling vast amounts of consumer goods, and the role that mass-market consumption and automotive technology had in introducing such conveniences to her. She may now ask questions about vehicle emissions and climate change, but only because of profound shifts elsewhere in the relevant system—not only changes in "natural" phenomena (which she may be aware of because of media coverage, but which she will not have experienced personally, since no one ever directly perceives "climate change"), but also in cultural and social patterns.

We already have a sharp contrast here between Level I and Level II. We humans, by definition, live in a world of technology.

We live in cultures of technological innovation, upon which we depend for our survival, and which conditions our behavior. At the first level, this dependence reflects our direct efforts to exercise our intent across an increasing breadth of human domains to accomplish certain tasks with great reliability, even as, at Level II, it enmeshes us in complex socio-technical systems of Kafkaesque incomprehensibility and capriciousness—systems that themselves demand continual technological updating in the process of adjusting to unfolding and unpredictable complexities. The airplane is experienced in all three ways: as an extraordinarily reliable piece of technology, as a component in an extraordinarily irritating transportation system, and as a major mechanism by which warfare and projections of military power are changed, tourism and its impacts on fragile, previously unreachable ecosystems are facilitated and disease vectors are distributed globally.

Imagine you are a fisherman sitting in a cod-fishing boat in Boston Harbor back in 1975, when there still were cod in Boston Harbor. (One of us, having dropped out of college in an effort to experience "the real world," really did this.[10]) Your boat is small, and the cod fishery is in decline, so you've had to switch your fishing gear from gill nets to tub-trawls—long lines with hundreds of baited hooks on them—in order to get a decent catch. The costs of bait and fuel are barely covered by the catch, and consequently you are being paid in sandwiches and all the cod you can eat. While relying on these basic technologies, you are within sight of Logan Airport, and planes are continually landing and taking off, every minute or so, with reliability and precision that make your own elemental struggles to bring in a few fish seem to be from a different world. One of the major differences between you in your boat and the jet soaring over you is how much of the system can be integrated directly into the technology. Your ability to succeed as a fisherman depends, of course, on some technologies (the boat, the nets, the tub-trawls), but the most important parts of the system—the fish, and the ecosystem that sustains them—are outside of your control. The jets, on the other hand, are elements of a technological system that internalizes almost everything that is necessary for their functioning. The design of the wings is such that the planes have no choice but to take off. The design of the jet engines provides tens of thousands of hours of reliable operation—and more recent generations of engines phone in to their maintenance departments if that reliability appears to be in question, so they can be inspected and fixed before the next flight. Not only that, but aircraft design is subject to continual, incremental improvement through, for example, the development of lighter, stronger alloys and composites, more precisely machined parts, and more aerodynamic designs. But when you are a fisherman, there is no engineering yourself out of the absence of fish—in fact, better engineering of fishing boats makes them more effective in reducing the stock of fish, thereby making the problem worse (a case in which Level I and Level II come into conflict, in that meeting Level I goals—build a boat that catches a lot of fish easily—conflicts with Level II goals—maintain a healthy population of fish over time, so that fishing can continue indefinitely).

Of course this isn't the whole story. It's not just that the planes are reliable, it's that the criterion of reliability for this system are completely obvious, and there is broad agreement that unreliability is completely unacceptable. People agree that the essence of a reliable airplane is one that gets you to your destination without crashing. This value is widely shared. And just about everyone agrees that, if you want to travel many miles in the shortest possible time, airplanes are the best option. So there's a sort of transparency to the effectiveness of airplanes that transcends ideological, religious, or philosophical differences. Not only that, but the Level II air transport system—for all its dysfunction and irrationality—depends utterly on the reliability of the Level I technology, the airplane itself. The system has to figure out how to maintain that reliability at all levels of operation, from designing, building, and maintaining aircraft to managing air traffic to detecting wind shear. The feedbacks from failure, and the incentives to learn and behave appropriately, are very clear and strong. In 2007 and 2008 U.S. commercial airlines flew 22 million scheduled flights, traveling a total of about 16 billion miles, with no fatal crashes,[11] a record that reflects continual learning and improvement as the size and complexity of the system grows. Meanwhile more and more travelers are complaining, and more and more people are refusing to fly because of the security and ticketing hassles, the charges for every incidental, and the airlines' inability to get their extraordinarily safe planes from point A to point B with any meaningful correspondence to posted arrival times.

Thus, our experience with flight is filled with contradiction at the interface of Levels I and II. We are powerless consumers, buffeted by the turbulence of organizational incompetence that seems to grow more pathological with the years, even as we can justifiably maintain a high level of confidence in the operation of the core technology. Yet not only do airlines and the producers of their physical technology platforms manage to keep their planes in the sky (more and more of them, in fact); they keep developing planes that are more energy efficient, have more range, and are more reliable, building a record of steady improvement—indeed, a record of progress. Very strange.

What can this have to do with transhumanism and human enhancement? Well, one thing to be said is that there was a time in the past when the idea of millions of humans flying around in the air would have been seen as a considerable enhancement over current capabilities, even if now it is not only mundane but largely aggravating. Given that many technologies aim in some way to create a human ability to do something better than could be done without that technology, we have already suggested that the whole human commitment to technological innovation could just be reframed as a commitment to the enhancement of human capabilities. There is certainly something to this position, but it is also the sort of irksome argument that some bioethicists and transhumanism advocates have used to make sweeping claims on behalf of technological promiscuity. On the other hand, it seems to us that the capacity of technologies to allow people to more effectively exercise some modest form of rationality in the world does in fact get us closer to the heart of the matter—to the issue of what it is likely to mean for people to be "enhanced."

We particularly want to call attention to the difference between Level I, where intent has a good chance of being translated into a desired outcome through the use of a technological artifact, and the larger, complex, often dysfunctional system of Level II, where the consequences of action become very difficult to map out, so that projecting intent becomes at best a matter of trial and error and at worse close to futile. Moreover, many institutions, including more abstract ones such as law and cultural patterns, have a tendency to focus on Level I—because it's simple, reliable, and easy to understand—and then to lock in on Level I solutions, with the result that they are not able to adjust when adverse Level II behaviors emerge. That's part of the reason fisheries collapse.

In some ways it is an understatement to suggest that our focus on different levels of technological effectiveness goes against the grain of much thinking about technology and society for the past 40 years or so. Classic works that helped to define this field, such as Jacques Ellul's *The Technological Society* and Langdon Winner's *Autonomous Technology*, looked at the ways in which societies restructure themselves to accommodate evolving technological systems and saw, for the most part, a ceding of human agency and authenticity and democracy to the demands of the system—Level II and nothing but Level II. The Level I functionality of the technologies themselves was a trick, a sleight of hand intended to distract us from what was going on at the higher level. And it's true: complex technological systems do make a mockery of the Enlightenment fantasies of rational control over our affairs and dominion over nature. The works of Ellul, Winner, Mumford, and others are suffused with an appropriate sense of befuddlement and resignation about what we can do to bring the technological systems that we create under our more direct and effective and democratic control. But an airplane is an impressive statement of bounded control in response to a particular challenge: You want to get from A to B safely, routinely, quickly. You even think this freedom is worth the aggravation of dealing with the institutions created to help you do that. Fine. We'll build you an airplane. Confusion arises not from the impressive essence of technological function, but because the analyses keep jumping from one level to another—from, say, the airplane level to the level of transportation networks as not just necessary for the operation of the airplane, but as instantiations of market capitalism and state power—without understanding that each level raises very different performance, management, and policy challenges even while nucleating around the same artifact, in this case our humble airplane.

Thus, when Winner complains about "reverse adaptation—the adjustment of human ends to match the character of [technological] means,"[12] to describe our subservience to technological systems, the question "compared to when?" kicks in. When was that golden age of harmony and agency and freedom from adjusting our ends to the available means? Here is a whiff of our old Cartesian mindset. Haven't humans always adjusted—"reverse adapted"—to their environment and available technologies? As the world gets more complex, of course, the systems dynamics may well become proportionately more difficult to even describe, let alone comprehend; however, this is true not only of technological systems, but also of institutional, social, economic, and other systems we live with. This point is not, we want to emphasize, an apology on behalf of complex technological systems, nor is it wallowing in technological optimism; rather, it is the suggestion that maybe there is more than one way to look at how technologies mediate people's lives—not just seeing the vast, hegemonic, unthinking Level II system in which technologies are embedded, but also the local, homely, quotidian Level I of actual use.

This is not mere quibbling or academic hair-splitting; it goes to the heart of any effort to understand the meaning of our world. Many environmentalists think that the automobile is a terrible technology, calling forth paved deserts, global climate change, obesity, and so forth. In this view, marketing has foisted SUVs on a willing public in spite of the environmental costs they entail. And who can deny that? But at the same time, automobiles are, in many cultures, icons of personal freedom and expression—not for nothing do societies that wish to oppress women forbid them from driving.[13] So are automobiles the institutionalization of environmental evil, or an expression of personal freedom? Both effects derive from a link between the Level I functionality that attracts people to use cars in the first place and Level II complexity, but the question is incoherent, for it conflates two different views of the technology system, two different sets of effects, and two unrelated questions of values.

A different example will allow us to further pursue the conundrums and contradictions created by the simultaneity of technological effectiveness and system complexity. We are impressed by vaccines as an illustration of many of the possibilities, conflicts, and difficult choices raised by transhumanism and the technological enhancement of humans, so let's use them as a way to push a little harder on this contrast between larger system and local use. Vaccines represent precisely the sort of internal, technological enhancement of human biology that seems at the core of the transhumanist agenda. We are introducing a foreign material into our bodies in order to stimulate a response by our immune system, and to enhance our resistance to a variety of infectious diseases as a result. In many ways, in fact, vaccines represent everything one could want in a human enhancement. Extended, in some cases even life-long immunity to a variety of diseases can be delivered through a process that takes seconds to administer and confers its benefits with remarkable consistency and reliability. A vaccine can be administered easily, by people with minimal training, in just about any setting. The effectiveness of most vaccines is quite apparent: smallpox, a truly horrific disease, has been eradicated, we have come close to eradicating polio, and vaccines have played an important role in the reduction of childhood disease and mortality in many parts of the world. Though vaccines have often stimulated opposition on moral grounds and because of concerns over risks, on the whole they are widely accepted and embraced.

A vaccine is thus an exemplary, bounded, Level I transhumanist technology. But the contrast between the effectiveness of vaccines and the chaos of the system within which they are administered may be even more conspicuous than for air transport. The U.S. health-care system, for example, has become an emblem of inefficiency, dysfunction, and inequity, yet most people manage to get necessary vaccinations and enjoy the protection that they confer. Even in countries with little public-health infrastructure, vaccines have often proved to be a powerfully effective intervention for improving health. Thus, vaccines strongly illustrate the dual realities of Level I technological effectiveness and Level II system complexity.[14]

But what makes vaccines more effective than other approaches to reducing the toll of infectious diseases? Let's explore this briefly by considering malaria, a disease for which no vaccine is yet available. One of the more conspicuous failures of modern science and technology policy has been the relative neglect of research on diseases such as malaria that especially afflict poor people living in poor countries. In the past decade or so, private philanthropy, sometimes in partnership with international organizations, national governments, and the private sector, has stepped in to try to reduce this imbalance. For example, research on malaria, and on malaria vaccines, has increased significantly. Yet the technical obstacles to developing a malaria vaccine are daunting, and it is not clear how long it will take for research to yield useful vaccines, or even if they are possible. In the meantime, malaria is killing at least a million people a year, most of them children, most of them in Africa.

In the absence of a vaccine, there are several major cooperative efforts under way to promote prevention strategies involving the distribution of insecticide-impregnated bed nets, limited indoor spraying of insecticides, and other measures. In many ways, insecticide-impregnated bed nets are a very appealing technology: inexpensive, low tech, simple to use. Where such nets are in wide use, the spread of malaria declines rapidly. Yet the first decade of concentrated effort to promote widespread net use in malarial Africa was widely deemed a failure, with the incidence of childhood malaria *increasing*. Reasons alleged for this failure include policy disagreements over appropriate distribution methods, the bureaucratic incompetence of the U.S. Agency for International Development, the cooperative Roll Back Malaria campaign, and other organizations, and simple lack of use by those who received nets.[15] Despite significant expenditures and effort, by 2005, after a decade of effort, only about 3 percent of African children in malarial regions were sleeping under insecticide-impregnated bed nets.[16]

In response to these disappointments, a policy consensus began to emerge in the mid-2000s around an approach called Integrated Vector Management. IVM combines bed nets with other interventions in a way that, according to the organization Africa Fighting Malaria, "means tailoring a variety of preventative interventions to local context."[17] As described by the World Health Organization, the characteristic features of IVM include "selection of methods based on knowledge of local vector biology . . . rational use of insecticides . . . and good management practices. The specific mix of these interventions depends on local factors, such as the type of mosquito and malaria parasite, climate, cost, and available resources."[18]

Very promising preliminary results of IVM in several countries have created a sense of optimism about the prospects for making real progress in combating malaria. A report by McKinsey & Company (commissioned by Roll Back Malaria and released to the public at the 2008 Davos World Economic Forum) suggested the following:

[A]n investment [in IVM] of approximately $2.2 billion a year for five years . . . can achieve full coverage of prevention and treatment measures in the 30 hardest-hit malaria endemic African countries, which together account for an estimated 90 percent of global malaria deaths and 90 percent of malaria cases in Africa. . . . Over five years, this effort is expected to:

- Save 3.5 million lives
- Prevent 672 million malaria cases
- Free up 427,000 hospital beds in sub-Saharan Africa
- Generate more than $80 billion in increased GDP for Africa.[19]

Such a radical reduction in malaria would be a magnificent achievement—true progress that we should all hope for. But it is hard to be optimistic about these predictions, for two very different reasons. Let's imagine that, instead of applying Integrated Vector Management to the prevention of malaria, we had a reasonably effective vaccine. What would be different?

Certain things might be the same. There would, no doubt, be controversies over appropriate policies for vaccine delivery, there would be organizational dysfunction at various levels, and a continued lack of adequate infrastructure for delivering health care in parts of Africa. In response to such hurdles, the essence of IVM, the key to its success, is "tailoring a variety of preventative interventions to local context." Yet surely the less tailoring that is necessary to achieve a desired result, the more likely success will be. To the extent that IVM depends on responding to local context, it also depends on managing and implementing knowledge and action in ways that are particular to that context. In any particular setting, the appropriate mix of interventions—"a combination of bed nets, medication, spraying, and environmental management"—must be determined, and the institutions and people responsible for delivering the interventions must act appropriately. IVM, that is, is a complex combination of activities that requires organizations to behave in particular ways in particular settings. And, crucially, no single activity embodies the essence of malaria prevention. In complex organizational settings, learning is particularly difficult because it is often not obvious how the lessons from one context might apply to another, and disagreement over competing approaches is common because the links from cause to effect are difficult to specify fully—numerous interpretations of success or failure may be plausible. People and their institutions aren't nearly as reliable, predictable, or replicable as good vaccines.

The main selling point of IVM—its sensitivity to context—is also its weakness. It isn't that current approaches to controlling malaria shouldn't be tailored to context—surely they must be. But tailoring to context is hard to do. A really effective intervention is one that, to the greatest possible extent, renders context irrelevant (or, to put it another way, internalizes context in the technology so worrying about context is not necessary). If a reasonably reliable malaria vaccine were to be developed, no doubt there would be challenges related to cost, manufacturing, and social acceptance of the vaccine. But much more of the intervention—the action that leads to the desired outcome—would be embodied in use of the technology itself. What you have to do to succeed is clear and always the same: vaccinate people. If people get vaccinated, they will, with high reliability, be immune, no matter where they live, how they live, or what they believe. People may or may not reliably use the bed nets they are given, but once they have been vaccinated the problem is solved—the locus of reliability shifts from the individual to the technology. Perhaps there will never be effective malaria vaccines, in which case IVM may well be the best path to reducing malaria in poor countries. But IVM is likely to be a very indirect path, one that does not always reach its goal. An effective vaccine would do the job better.

The process of delivering a vaccination is what we might call a shop-floor activity—an activity whose effectiveness depends little if at all on the larger organizational setting.[20] Here we are making a subtle but important distinction between the vaccine itself (a Level I *technology*) and the process of administering the technology (a Level II *activity*). As with the airplanes in the dysfunctional transportation system, most of the cause-and-effect elements directly relevant to solving the problem are localized in one particular technology whose performance can be easily measured and is largely insensitive to the surrounding institutional context, and just about everyone agrees on what counts as success. Level II activities can be organized to achieve this success, because its attributes are obvious, and feedbacks from failure are clear (recall that the dysfunctional air transport system can nonetheless organize around the achievement of airline safety). Solving a problem is hard when you don't have a way to condense the major cause-and-effect elements of the solution into a particular technology or routine that can be administered on the shop floor—that is, at the Level I site where the action takes place, and where the results of the action can be experienced. By our definition, a Level I technology is simple (even if it involves a sophisticated feat of engineering, like a modern jet airliner) in that it can take advantage of capturing both the cause and effect involved in a particular task or activity; but both Level II and Level III involve complex adaptive systems in which causes and effects are difficult if not impossible to isolate.

Consider an example from our experience that literally involves shop floors: the use of toxic chemicals in a manufacturing process. One way to manage such a risk to employees is to provide appropriate protective equipment and mandate its use. But protective equipment can break down, be misplaced, or simply not used. (Employees often dislike protective equipment because it is uncomfortable and makes their job harder; managers may not be as strict in requiring it as they should be, because it may impede productivity.) Alternatively, one can redesign the manufacturing process to use less toxic materials in the first place. The former (an "IVM approach") is never as effective as the latter (a "vaccine approach"). This intuitive understanding is embodied in the well-known occupational health and safety heuristic known as the "hierarchy of control," a chain of potential response mechanisms, ranging from the most efficient to the least efficient:

elimination of the risk

substitution of less risky alternatives

isolation of the risk to reduce potential exposure or impact

implementation of engineering controls

implementation of administrative controls

use of personal protective equipment.

Note that the first two options operate at Level I: the risk is designed out of the technology. The other options increasingly rely on institutional and social systems rather than technological design. This method of understanding risk management is not just academic. One of us was at one time the executive in charge of environment, health, and safety for a large firm, and can testify to the validity of this concept: the more responsibility for safety you can transfer from human and institutional decision processes to technologies themselves, the safer the system will be, all else equal.

Of course many important problems can't be technologically embodied in this way. It is not just that sometimes the technology can't be developed with our current knowledge base (as in the case of a malaria vaccine); it is also that the ability to internalize goals in technology depends crucially on the goal in which one is interested. The report on IVM we mentioned earlier predicts that an investment of $10 billion will lead not only to reduced malaria cases and mortality in Africa, but also to enhanced wealth creation—$80 billion over five years, to be exact. One should expect healthier people to be better able to contribute productively to an economy. But there are so many other confounding and contributing factors, including education levels, environmental conditions, quality of governance, and the global trade situation, that any prediction based on links between changes in malaria incidence and changes in wealth creation is at best a vague and hopeful guess about behavior in a complex system. This difficulty is nicely illustrated by the work of Peter Brown,[21] an anthropologist who tested the hypothesis that malaria was blocking economic development on the island of Sardinia in the period after World War II. He concluded that the "macroparasitism" of landowners drained 30 percent of the production capacity from peasants in the form of rents, while the "microparasitism" of malaria accounted for a reduction of less than 10 percent in their gross production. And here we shouldn't expect a vaccine to do any better than bed nets; here the goal—creating wealth—cannot be captured and internalized by a particular technology. In fact, if creating wealth is your goal, there may be much better routes to progress than curing malaria—for example, changing patterns of land tenure, or improving levels of education. But of course these goals are themselves very hard to make progress on.

Understanding a technology is not just a process of observing something "out there"; it is an integrated result of a query, a set of artifacts, and elements of social, economic, psychological, and cultural context, called forth as a whole. Each query implicitly identifies certain elements of the underlying system as relevant and ignores others—a process that is perfectly legitimate until one extends one's judgment or analysis beyond the boundaries that are also implicit in the query, when it might break down. Such breakdowns are common in much of the literature on transhumanism. In chapter 2, for example, we suggested that even if a cognitive enhancement works at an individual level, that doesn't mean that humans, as social animals, would be better off in aggregate, even by utilitarian measures. This scale-up problem can now be seen as an example of a general error, where the individual results of employing a Level I technology on the shop floor are assessed in a manner that might be perfectly correct but then are extended to a higher level of system complexity with no understanding that an important boundary to the validity of the analysis was being jumped.

Now let's push a little farther into transhumanist territory.

Cochlear implants are electronic devices that provide deaf people with a sense of hearing by direct stimulation of auditory nerves. Unlike hearing aids, which merely amplify sound, implants can give profoundly deaf and severely hearing-impaired people the ability to sense and interpret sound, including speech. First tested in the early 1960s, cochlear implants were approved for use in the United States by the Food and Drug Administration in 1985. By 2009, about 190,000 devices had been implanted worldwide, most of them after 2000. New technologies promise even more dramatic capabilities.

Cochlear implants have been opposed on the ground that deafness is an aspect of human diversity, not a deficiency to be corrected. From this perspective, deaf culture is as rich and worthy of protection as any other distinctive culture, and deaf people don't need to be "enhanced." Deaf people use sign language to communicate richly with one another and with hearing people who also sign. Obstacles to complete and equal participation of deaf people in a society dominated by hearing people are a reflection of institutionalized injustice, not of the attributes of deaf people. The appropriate ethical goals, therefore, are to eliminate obstacles to full participation in society by deaf people and to promote complete acceptance of deaf culture. Cochlear implants are a threat to these goals and to the sustainability of deaf culture, and should therefore be resisted. In support of this position, at least one deaf couple sought out a sperm donor with a long history of family deafness to increase the likelihood that their child would also be deaf.

The reasons why people choose cochlear implants are apparent. The main ethical subtlety lies in the fact that many devices are implanted in young children who cannot consent to the operation or to the enhancement (hardly a unique situation), and in such cases it is presumably hearing people for the most part, rather than deaf people, who are making the decision. One can certainly argue that opposition to cochlear implants is based on a generous vision of social justice and equity. But justice and equity can be defined and served in different ways. If the goal is to create a society in which deaf people have all the opportunities and benefits that are available to hearing people, and if two paths are open—the struggle for the complete and equal rights and access of those without hearing, and the widespread adoption of cochlear implants—it is likely that the first path will be much more difficult, uncertain, frustrating, and protracted than the second. As in the case of malaria, one option pushes the complexity of the larger system to the background by embodying most of the relevant cause and effect of the immediate problem in a technological solution—and in so doing it radically reduces the political and organizational challenges involved in making progress toward the goal. This is classic Level I technology. But to some people involved in the dialog, the sharp boundary to the problem defined in this approach is inappropriate, in that it leaves out important considerations arising from the psychological, social, and cultural context. Opposition to cochlear implants thus engages the technology as a Level II system, and thus implicates much more complexity, difficulty, and uncertainty than are found on the shop floor.

Alert readers may have sensed that we are beginning to get into a little trouble on the ends-means front. Is the goal to ensure that deaf people can participate as fully and equally as possible in society? Or is it to ensure that deaf culture is sustained and fully accepted and included in a vibrant and diverse society? It could be true that a society that fully embraced deaf culture as an expression of human diversity would be a more just and equitable society than one that radically reduced deafness through the widespread use of cochlear implants. Indeed, we think that probably is the case. But modifying deaf individuals technologically so that they can participate in hearing culture is simply an easier task—much easier—than stimulating the political and behavioral changes that would get society to fully include, nurture, and sustain deaf culture and, in the process, make implants irrelevant. The dilemma exemplifies Langdon Winner's complaint about adjusting our ends to match the character of our technological means. But such adjustments also allow us to act more effectively. Here we are not making an ethical argument for or against cochlear implants; rather, we are making an observation about the likely success of different paths of social change, and a weak prediction about the choices that people on the whole are therefore likely to make as they seek to achieve particular goals.[22]

So we have a situation pitting the ease of a reliable technological intervention against the hard and unpredictable slog of political struggle. What gives the technological option the fundamental political advantage is its functionality, its effectiveness on the shop floor—if it didn't do what it claimed to do, its competitive advantage against the non-technological option would be harder to sustain. The Level I effectiveness of the implant itself has a political meaning, a built-in political impetus; and the implant attracts various constituencies that want to get a certain thing done, because it does so more reliably than other means to a related end. The dilemma is particularly disturbing because it rubs our noses in the possibility that what seems like the right thing to do—work for a more tolerant and inclusive society—may in the end be a less beneficial path to follow than the use of the technological intervention—not because it wouldn't be better to have a society that didn't need the technological fix, but because the fix is so much more effective, more reliable, and more predictable than the political path toward progress. This does, however, suggest a caution regarding any technological fix: Be careful to understand the Level I goal that the fix actually addresses (e.g., reducing deafness or preventing malaria), and distinguish it from more complex Level II goals (e.g., creating a more tolerant and multicultural society that embraces deafness; achieving more rapid economic growth in malarial regions) that the fix may not address, or might even reduce pressure to achieve (or with which it might even conflict).

Let's pursue this tension one uncomfortable step further. One of the most conspicuous sites for technological enhancement of humans is the process of childbirth. The influence of technology on the birthing process has been pervasive and profound and seems likely to become more so. One may also feel it has been alienating and dehumanizing. At the same time, the industrialization of childbirth, through the application of technologies such as labor-inducing drugs and heart monitors and standardized procedures like the Caesarean section and the Apgar score, has also made childbirth far more reliable and predictable than it once was. In rich countries, infant mortality at childbirth has declined from several hundred per thousand in the eighteenth century to ten or fewer per thousand today, and maternal mortality has declined from as high as 10 percent as recently as the late nineteenth century to less than one in 10,000 today.[23] Looking at these trends in conjunction with the rise of assisted reproductive technologies such as *in vitro* fertilization and the increasing ability to nurture radically premature babies one can imagine that we may be on a course toward pregnancies that are completely technologically mediated, perhaps even outside the womb, to deliver utter reliability in the outcomes of childbirth. The benefits of this historical trajectory are undeniable, even as the continual intrusion of technology into pregnancy and childbirth may reasonably offend our sense of what is appropriately human. This offense may seem to be magnified when we think about a related matter: the entrenched inequities in birth outcomes in the United States. For example, the rate of infant mortality is about twice as high among African-Americans as among white Americans; and the overall rates of U.S. infant mortality have long been unconscionably high relative to other rich countries, mirroring America's greater levels of socioeconomic disparity. So we pursue all this technological change in our affluent society, but meanwhile we can't even do what is necessary to ensure that poor and minority babies have the same chance of surviving as white and well-to-do babies.

But there are, it turns out, two twists to this tale. First, over the last few decades, infant mortality rates among poor and minority babies in the United States have declined at about the same rate as among the babies of more well-to-do parents. The disparities remain distressingly resistant to change, but the absolute outcomes have improved for everyone. These declines appear to be due almost entirely to Level I technologies—shop-floor interventions in the delivery room that have provided tremendous benefits to poor and well-off alike. Second, there have been substantial efforts to address the inequity at higher levels of complexity, and they have largely failed. More than 40 years of policies aimed at increasing the quality of prenatal and maternal health care and nutrition among poor women in the United States through Medicaid and other programs have had little or no clear positive effect on birth outcomes. These worthy efforts turn out not to have narrowed the mortality disparities.[24] The specific reasons for the lack of progress are, of course, debated among the experts. The causes of high infant mortality rates among poor people are complex and are deeply embedded in broader problems of socioeconomic inequity that continue to resist political solution and effective policy intervention.

Obviously we are not arguing against engaging in the hard political battle for greater socioeconomic equity in our society; the growing concentration of wealth in our already wealthy country and in the world is a continuing moral affront. Rather, our point again is that when the essence of a problem is amenable to a technological intervention, real progress can sometimes be made very rapidly with a Level I technology, whereas political paths to solving a bigger, underlying problem are likely to be much slower and less satisfactory, even if they engage the same underlying set of artifacts. This is what we are seeing in the case of infant mortality.

Yet the technological path may seem less ethically satisfactory than the political path, because it leaves intact the underlying social failures that contribute to the inequity. Again, we adjust our ends to suit the available means, and this may create a reasonable sense that the technological path provides us with an excuse for not taking the political path—a sense that the available means are distracting us from the more important end, from doing what is right, which is to solve the problem by making society better, by reducing inequality, rather than by isolating the problem from its social context through a technological fix.

And this again leads us to emphasize the incoherence and wrong-headedness of much of the dialog around transhumanism. Technological enhancement of humans is not going to help us confront the most fundamental political challenges faced by our societies, challenges underlain by a combination of value conflict and uncertainty about the future consequences of our actions. Analyses that frame technologies in a Level I context—as specific means to identifiable and simple ends, such as better memory, less anxiety, or more focused attention—simply cannot be plausibly extended to imply that those technologies represent solutions to more complex social and cultural phenomenon. It's a category mistake. The stubbornly persistent socioeconomic inequities that continue to fester in the United States, and that underlie our poor performance on infant mortality, are one example of a Level II problem that is not amenable to Level I technological solution. Enhancing concentration or memory will not make us more moral, caring people. But here we assert a complementary point: There is *no* easy path to addressing such fundamentally political challenges, but technologies can sometimes help us find a shortcut to dealing with some of the particular consequences of these challenges.

Accordingly, we simply don't buy into the framing of transhumanism as offering a choice between two alternative technological futures, one essentially utopian and the other essentially dystopian. We suggest instead that there is a scale of experience where one doesn't have to give up one's sophistication about the complexity of the world (and its technology systems) to accept the possibility of modest yet encouraging technological progress on the shop floor. When vaccines work, they are good things because they reduce the incidence of the targeted disease. This progress derives from our innate capacity and our apparent compulsion as a species to innovate technologically—to take certain types of problems and capture much of what is difficult about them in physical artifacts that allow us to get around those difficulties.

In making this kind of progress, we are creating a domain of increasing control related to solving a particular problem, even as we are also feeding into the complexity of socio-technical systems—Level II—whose comprehensibility continually slips from our grasp, and often presents us with new types of problems. This seems to us a fundamental dilemma of the techno-human condition, one that requires continual, balanced attentiveness. Technology is no more the cure for politics than politics is the cure for technology; they need each other, can benefit from each other, and evolve with each other, and we are forever stuck with both. Yet we can also recognize and appreciate that there is something different and special about technology—something that, under certain circumstances, allows us to act in the world with much greater effectiveness than we could otherwise achieve. In fact, we would go so far as to say that the greatest source of reliable action in human affairs is not our institutions, our cultures, or our norms, but our inventions. Any approach to solving the many vexing challenges that face the world today must accommodate this fundamental though uncomfortable reality.

The transhumanism debate can now be seen in a new and different light. Different sides of the debate are arguing about different levels of technology without realizing it. The proponents tend to discuss Level I technologies—specific enhancements, administered and experienced on the shop floor, that address specific and identifiable goals, such as improved cognitive performance. The opponents voice concern about major changes in current system states by focusing on complexity and on normative values—in other words, their discussion is at Level II (if not Level III). Proponents and opponents both err in failing to recognize that both positions may be at once valid yet incommensurable. Any technology of more than trivial import exhibits behaviors at Levels I and II (and beyond, as we will discuss next), and these behaviors are unavoidable and symbiotic. But more profoundly, by relying on simplistic, anachronistic, and contradictory conceptual frameworks, both sides reinforce concepts and frameworks—Enlightenment certainties—that are not capable of engaging the radical technological transitions that we humans are continually creating. It is to these transitions that we now turn.

4

Level III Technology: Radical Contingency in Earth Systems

We have explored two levels of technology. At the shop-floor level, we can see much of the cause-and-effect chain necessary to meet specific and well-defined social goals: a vaccine prevents a particular disease, or a well-designed manufacturing process eliminates the use of toxic chemicals (and thus workers' potential exposure to them). At the second level, technology is a networked social and cultural phenomenon; any particular technology functions in a broader context that can be complicated, messy, and far less predictable or understandable than what happens at the shop-floor level. Still, we are generally familiar with this second level; we talk about transportation and information and communication technology (ICT) networks, about retailing and food-supply systems, and about energy, water, and power systems. We can *see* what the technologies do, and we can recognize what is in the system and what is not, even though acting to achieve a particular intended outcome is often difficult because the internal system behavior is too complicated to predict.

But there is a third level that we are not so familiar with, a level at which technology is best understood as an *Earth system*—that is, a complex, constantly changing and adapting system in which human, built, and natural elements interact in ways that produce emergent behaviors which may be difficult to perceive, much less understand and manage. And at this level technology is not just a complicated network that is bothersome in its inability to behave as we would prefer it to, such as the air transportation system or the health-care system; rather, it is a transformative wave that towers above us, ready to crash down—not just an organizational or political or cultural force, but an existential force. At this level, it's not just that you can't handle the truth; it's that you can't even grasp it; it's too complex to be given in forms (ideology, scientific models, traditional values) that you can process. You may well disbelieve this. That's why, later in this chapter, we will use the railroad, as mundane a technology as you could imagine, to make our point. Mundane to you, sure——but you come after, not before, the "railroad singularity."[1]

As we've said, these three levels of technology—like most human-created categories—have fuzzy boundaries and some unavoidable arbitrariness. We make no apology for this. Indeed, when we consider technology in this way the failings of the cultural construct of transhumanism comes into sharper focus, while the hard categories of Enlightenment thinking—mind/matter; natural/artificial; individual/social—get increasingly fuzzy.[2] In particular, the mismatch between the reductionist and directed rationality of the Enlightenment that is built into Level I technology, and the challenging and highly complex environment of Earth systems, requires for its comprehension (as limited as our capacity to comprehend may be) nothing less than a new frame of reference for understanding and for action: a reinvention of the Enlightenment. We are suggesting that without a new and difficult ascension to a rationality suitable for a world in which "all that is solid melts into air" (Karl Marx's words) we forfeit our already tenuous hold on responsibility and on ethics, and perhaps even our increasingly questionable claim to be considered sentient beings. To put it slightly differently, the world we are making through our own choices and inventions is a world that neutralizes and even mocks our existing commitments to rationality, comprehension, and a meaningful link between action and consequence. Either we accept that we are impotent brutes living way beyond our means because of the technological house of cards we occupy or we search for a different set of links to connect our highest ideals to the reality we keep reconstructing.

We are now ready to see why technology, the idea of progress, and other gestalts swirling around transhumanism often devolve into fruitless dichotomized conflict. The simple answer is that, having failed to parse technological levels in a useful manner, we confuse them in ways that ensure misunderstanding. Conflict arises because interacting complex systems express different behaviors at different levels; indeed, that is the idea behind the concept of emergent behavior. Moreover, different levels of a system need not address the same goals, values, and questions. At the shop-floor level, technologies captured in artifacts often produce progress toward specified goals (that's why they are adopted). At Level II, the goals may still be visible, but because of the qualitatively different complexity of these networks (e.g., the policy, cultural, and economic networks with which the shop-floor technology is now coupled) performance is usually patchy (think of the delivery of medical services in Africa or in the United States, or air transportation as an overall experience for the traveler). Progress, if it comes, is an outcome of embedding more of the system in the technology itself, or, failing that, of the much less reliable process of political action. Finally, at Level III, technology systems need to be understood as transformative Earth systems. Academics (especially those debating transhumanism) generally have not dealt with Level III, because it is beyond any particular disciplinary or intellectual structure, and is characterized not just by complexity, but by radical contingency: the values, frameworks, and cultural constructs on which we rely are undermined by the technology they have made possible, and prediction and even judgment become dependent on a context that is always shifting, and on meanings that are never fixed. All things solid melt into air, and the usual response is to don disciplinary blinkers and define the problem away, or to turn the problem over to computer modelers in the belief that numbers—any numbers—will make it comprehensible and manageable, or to adopt a fatuous collapse into absolute relativism or dogmatism. We reject these responses.

But note well: at Level III, there are no unifying goals. With a transport network or a health system, the goal remains immanent in the definition of the system—it's the *transport* system, or the *health* system. With an Earth system, there are no agreed upon, universally valid goals; there is only non-directed, non-predictable evolution. Thus much of the confusion surrounding transhumanism is in fact confusion about how to perceive, think, and act in a world in which, by our actions and the cumulative evolution of our technologies, we have launched ourselves into Level III with only the tools of our vaunted Enlightenment—a Level I sophistication that has been characterized by the Indian anthropologist Shiv Visvanathan (2002) as "moral infantilism."

The Anthropogenic Earth

We live in a world dominated by one species and the activities and products characteristic of that species, from automobiles to cities to the creation of new cyberspaces. It is a world in which the critical dynamics of major natural systems, be they atmospheric, biological, or radiative, increasingly bear the cultural, economic, and technological imprint of the human. We cannot, in a reasonable amount of space, begin to weave an understanding of the complex adaptive systems that increasingly characterize this anthropogenic planet, but a small set of examples might provide a glimpse of what we have already wrought.[3]

Let us begin with a physically fundamental example. Every planetary body has a characteristic spectrum of radiation emissions that depends on its composition and its heat. Earth's spectrum, however, is not just a matter of reflections from clouds, emitted infrared radiation, and the like. It also includes television and radio broadcasts and leakage from all sorts of technologies. Remember those pictures of Earth from space at night, and the electric lights spread over North America, Europe, and Asia. In the Anthropocene, perhaps the most fundamental physical manifestation of our planet in the universe, its radiation spectrum, carries our signature.

Or consider a more topical example. Virtually everyone is aware of global climate change, which vies with terrorism for billing as the top existential catastrophe. Stand away from the Kyoto Protocol process and the surrounding hysterics pro and con, however, and take a longer perspective. What the climate change negotiation process taken as a whole represents, fitful and *ad hoc* as it is, is the dawning of a realization that, regardless of what results from current international negotiations, our species will be engaged in a dialog with our climate, our atmospheric chemistry and physics, and the carbon cycle so long as we exist at anywhere near our current numbers on the planet. This is not a problem, it's a condition. We can change— more likely, redistribute—some of our impacts on these complicated and interrelated systems, but we will not reduce the human influence. Moreover, these particular perturbations are not isolated phenomena but just one way to perceive the evolving behavior of interconnected global systems. A population of some 7 billion humans, each seeking a better life and thirsting for technologies used and perceived at the shop-floor level of complexity, ensures that our overall role in global systems will increase unless there is some sort of population crash. And be careful if you wish for this under your breath, for such a catastrophe, whether from nuclear winter, terrorism and response, ecosystem collapse, or some other source, would create havoc among all systems, human, natural, and built.

Another topical example is provided by the current "crisis in biodiversity," as human activity causes extinction to increase dramatically.[4] On the one hand, ecologists may be justifiably concerned about whether key ecosystems—the depleted wetlands surrounding New Orleans, for example—are able to fulfill the functions upon which our societies depend (protecting a city from hurricane surges, in this case). Yet from another perspective, biodiversity is fast becoming the next technological design space. Scientists and engineers have begun the project of understanding and designing new forms of life. These efforts, from genetics to agricultural science, have coalesced into a new field—synthetic biology—that merges engineering with biology by, among other things, beginning to create standard biological components that can be mixed and matched in organisms to provide desired functions. This will allow researchers to treat biological pathways as if they were components or circuits, and to create organisms from scratch, and to go beyond existing biological systems by creating forms of life based on genetic codes not found in the wild. The Massachusetts Institute of Technology has established a Registry of Standard Biological Parts ("BioBricks") that can be ordered and plugged into cells, much like electronic components that can be plugged into circuits. The Intercollegiate Genetically Engineered Machine (iGEM) competition held at MIT in November of 2005 attracted 17 teams, whose designs included "bacterial Etch-a-Sketches," photosensitive T shirts, and bacterial photography systems, thermometers, and sensors. A number of viruses have been assembled from scratch, including the polio virus and the influenza virus that caused the 1918 pandemic. (Scientists trumpeted and defended the latter achievement,[5] while two leading techno-visionaries, Ray Kurzweil and Bill Joy, pronounced the feat "extremely foolish" on the op-ed page of the *New York Times*.[6]) In 2010, Craig Venter, another techno-visionary, built a synthetic genome that was able to support reproduction when implanted in a cell. Other researchers have engineered the genes of *Escherichia coli* to incorporate a twenty-first amino acid, opening up an option space for design of organisms that has been unavailable to evolved biological systems for billions of years, since evolved biology is locked into the usual twenty amino acids (or at least was until human intelligence evolved to the point where new chemical bases for life could be created). Commercialization of these biotechnologies continues to accelerate, led by the introduction in agriculture of genetically modified organism (GMO) technologies—corn, soybeans, cotton, and other crops that are altered to achieve a variety of specific functions, such as increased resistance to pests or herbicides. But GMO technology extends far beyond agriculture, and biotechnology patent filings in OECD countries continue to rise sharply.

Synthetic biology provides a good illustration of some of the features of technological evolution at the scale of Earth systems. First, synthetic biology does not just reconfigure the biological sciences; rather, biodiversity becomes a product of design choices, and of industrial and political imperatives (security issues, for example), rather than a product of evolutionary pressures. More broadly, the behavior and the structure of biological systems increasingly become functions of human dynamics and systems, so that understanding biological systems increasingly requires an understanding of the relevant human systems. In short, biology, anthropology, and political science are converging.

As the tools of synthetic biology are used to address such shop-floor problems as curing diseases, boosting agricultural production, or generating energy, the consequences at the scale of the anthropogenic Earth will be essentially unpredictable.[7] This unpredictability arises from not only the complexity of the biosphere's evolution but also the increasing impact on biological systems of the contingency that characterizes human systems. For example humans have often chosen to preserve species that they find culturally or aesthetically appealing—such charismatic megafauna as pandas, tigers, and whales. Many, many other species go extinct because they are only insects, or plants, or ugly, or unknown; a few, such as smallpox, go extinct because humans detest and fear them (with the important proviso that, in the age of biotech, extinction—at least for viruses and bacteria—is not forever. Find the right mad scientists and they will rebuild smallpox for you if you want it and can pay for it.).

Perhaps the best way to understand the magnitude of Level III technological complexity is to imagine yourself as an alien seeing our planet for the first time. Not only are you surprised by the amount of biomass dedicated to the activities of one species; you also realize that Earth's surface—the cities, the conversion of huge regions such as the Argentine pampas and the American Midwest to agriculture, the transport and information networks, the high energy throughput—bears eloquent testimony to the planet as monoculture. We raise one crop on it—ourselves—and the other elements of the planet, from movement of sand and gravel to management of the hydrologic cycle, are increasingly yoked to that task. But it isn't just that our technologies construct a human Earth, and it isn't just that our technologies are ever more powerful means for integrating previously natural systems and human systems, because the human itself is part of what we are changing. These are not the sorts of changes over which transhumanists palpitate, for changing particular elements of cognition at the shop-floor level—say, by increasing memory or concentration or attention span with pharmaceuticals—is not the same thing as the consequent yet unpredictable changes in human institutions, behaviors, social structures, and the built and natural systems with which they are integrated. "The human," however one chooses to define it, is increasingly shaped by our technologies in a complex feedback process that is accelerating dramatically. To paraphrase Marx: We do indeed design the human, but we do not design the human as we intend. Indeed, if we are to comprehend, let alone move toward grappling with, the world we are continually remaking, we must get beyond the idea that we are imposing our intent, our purpose, on the future. Religion may be the opiate of the masses, but "cause and effect" is the opiate of the rational elite.

Technology and Creation

Any meaningful discussion of technology in the age of the anthropogenic Earth must emphasize the transformative role of technology at Level III, the level of Earth systems. At this level, technology is always coupled to other Earth systems, including other technologies, and this introduces dynamical behavior that is highly complex. (The ongoing games involving oil prices, biofuels, climate change, the automotive culture and the global economy should convince anyone of that.) But complexity is the least of it. Technologies destabilize the world, changing cultures, worldviews, power relationships, and ethical, moral, and theological systems.

Consider the railroad. In the middle 1800s it was not just the most impressive machinery most people ever saw; it was a sociocultural juggernaut. The world before the railroad differed profoundly from the world after. Things that people had regarded, culturally and psychologically, as foundational—their sense of time, for example, or their sense of nature[8]—were first rendered contingent, then swept away.[9]

As an integrated technology network on a regional scale, railroads required a uniform, precise system of time, and the need for "industrial time" thus co-evolved with its associated culture. We are so accustomed to "time" that we never consider how essentially arbitrary our system is; before railroads, however, things were quite different. Local times were isolated and charmingly idiosyncratic. For example, London time was four minutes ahead of Reading time, more than seven minutes ahead of Cirencester time, and 14 minutes ahead of Bridgewater time.

Alan Beattie (2009) points out that there were more than 200 different local times in the United States as late as the 1850s, at which point American railroads were using approximately 80 local times, a situation that became untenable as rail traffic and the speed of trains increased. Moreover, the adaptation to uniform systems of time was not smooth. For a considerable period, each train company in the United States had its own time, so stations serving several train companies had a clock for each company, each keeping a different time. (Buffalo had three such clocks at one point, Pittsburgh six.) Regional standard time did not gain legal recognition in the United States until 1918. In Prussia, the most militarized state in Europe and the one with the most advanced military technology and strategy, the General Staff of the military pushed for a standard time scheme to facilitate planning and operations, because their plans, operations, and strategies depended so heavily on the new railroad technology.

Major technology systems do not stand alone. Personal computers require high-quality power and software; automobiles require gasoline-delivery infrastructures; industrial-scale agriculture requires fertilizer, pesticides, and an efficient transport infrastructure. For their part, railroads created the need for, and co-evolved with, national-scale communications systems in the form of telegraph technology. This complementary technology system was both coextensive with rail networks (often laid along the same rights of way) and a necessary coordination mechanism for creation and operation of regional integrated rail systems. Networked technology systems that must be controlled to be effective, such as railroads cannot, after all, exist without mechanisms that ensure coordination of function throughout the network; in turn, this requires stabilization of time, and a communication mechanism of the same scale as the network, so that information regarding the current state of the network, directions and decisions regarding future actions, and responses and verification can be constantly communicated to ensure safe and smooth operations. Analogous functionality is built into the chips inside your computer, for the same reasons.

Railroad systems increased the scale of industrial operations significantly, and thus destabilized previous financial and managerial models. Railroad firms helped shape modern managerial capitalism because—unlike the factory system, which had only required a division of labor among factory workers—the scale of railroad enterprises required a division of labor at the management level, with modern and professional accounting, planning, human resources, and administrative systems. Moreover, the same scale issues were reflected in the role railroad firms played in the co-evolution of modern capital and financial markets. The earlier factory system was supported initially by aristocrats and landowners, and subsequently by factory owners using their own capital, an essentially individualistic financial structure—capital requirements were scaled by the size of the factories that individual capitalists built. Such fragmented and individual efforts were nowhere near adequate to support the huge capital requirements and geographic reach of railroad firms. By the 1840s, railroad construction was the most important stimulus to industrial growth in Western Europe.

Railroads did not just change institutions; they also transformed landscapes, both physically and psychologically. Chicago existed, and structured the Midwest economically, physically, and environmentally, because of railroads.[10] Psychologically, railroad technology did not just extend but obliterated the sense of place and rhythm that previous transportation technologies, such as horse-drawn carriages, or canal barges, had encouraged. Consider the psychological dislocation expressed by Heinrich Heine as he reflected on the opening of new rail lines across France in 1843[11]:

What changes must now occur, in our way of looking at things, in our notions! Even the elementary concepts of time and space have begun to vacillate. Space is killed by the railways, and we are left with time alone. . . . Now you can travel to Orleans in four and a half hours, and it takes no longer to get to Rouen. Just imagine what will happen when the lines to Belgium and Germany are completed and connected up with their railways! I feel as if the mountains and forests of all countries were advancing on Paris. Even now, I can smell the German linden trees; the North Sea's breakers are rolling against my door.

We moderns sometimes flatter ourselves that space-time compression is a uniquely modern ailment.[12] But in the nineteenth century the railroad radically changed, for the first time, the equation connecting distance to time. In so doing, the railroad powerfully and irrevocably reframed the relationship between human psychology and perception, and the external environment. The railroad thus reveals itself as a powerful transhumanist technology. Who knew?

But the railroad did more than (substantially) create modern industrial capitalism, the modern firm, the modern communication network, the modern urban landscape, and the modern sense of time. (By "create," of course, we mean "significantly force the co-evolution of.") Particularly in the United States, the railroad became a symbol of national power, and, more subtly, instantiated and validated the American integration of religion, morality, and technology. Ralph Waldo Emerson, Walt Whitman, and Daniel Webster were among those who—in an unconscious recapitulation of language and powerful cultural memes that we saw in Bacon centuries earlier and an ocean away, and now hear again in transhumanism—viewed railroads as evidence of human ascension to godlike power. In the early 1800s the Western Railroad in Massachusetts urged ministers to "take an early opportunity to deliver a Discourse on the Moral effect of Rail-Roads in our wide extended country." Similar language was, of course, used by those (not insignificant in number) who viewed railroads as Satanic, such as a school board in Ohio in 1828: "If God had designed that His intelligent creatures should travel at the frightful speed of 15 miles an hour by steam, He would have foretold it through His holy prophets. It is a device of Satan to lead immortal souls down to Hell."[13] The railroads initiated a coupling between religion and the technological sublime that was an important aspect of American exceptionalism, still evident today in arguments over transhumanism.

Not coincidentally, railroads also transformed military power relationships and strategy. Many Americans might point to the use of railroads by the North in the American Civil War as an example of competitive military advantage, but a far more interesting example is the rise of Prussia. After the 1815 Congress of Vienna, Prussia was a minor state, one among many littering Central Europe at the time; worse yet, it was split, with its two pieces separated by the independent states of Hanover and Hesse-Kassel, and it was surrounded by strong and powerful bullies: France, Russia, Austria. But Prussia had one thing no one else had: Helmuth von Moltke, the Prussian Chief of Staff. Von Moltke, in turn, understood the power of railroads. And when von Moltke crushed the liberal uprisings of 1848 in Prussia by rushing troops from city to city by rail, others in the Prussian military understood the power of railroads as well. A special Prussian Railway Fund was set up to build lines that were not economically viable but were important from a military perspective. All Prussian freight cars were designed to be able to carry soldiers, horses, and military equipment. The small Prussian army was organized into regiments, each of which was assigned to a specific railhead where it would assemble when mobilized. Railroads became the visible sinews of a military machine such as Europe had never seen. It all paid off in the battle of Königgrätz in 1866: the Prussians stunned Austria (a major European power), in part by managing to transport 197,000 men and 55,000 horses to the front using railroads—a feat the Austrians had not imagined possible. Of course, Prussian innovations in weaponry (especially the needle gun), and in training were also critical; as is always the case with complex systems, many factors came together in unpredictable ways. Königgrätz marked the end of the Austrian Empire and the emergence of Prussia as a European power, even though it lacked the economy, the population, and the geographical advantages of nations such as France or England.[14]

But let's follow this story a little further, for it is also a cautionary tale. In the early twentieth century, German military planners, building on the lessons of Prussia's earlier success using the railroad for military advantage, devised a strategy—the Schlieffen Plan—that would enable them to fight a war on two fronts. First they would achieve a rapid strategic success on one front, then use the railroad to transfer their troops to the second front while that opponent was still mobilizing. This was the German plan in World War I: avoid a war on two fronts, which the Germans recognized they probably couldn't win, by quickly defeating the French and then rushing those troops by rail to face the Russians, who the Germans (and everyone else) believed would take a long time to mobilize. But the unexpectedly vigorous French defense in the Battle of the Marne, and the fact that the Russians mobilized far more quickly than anyone expected, led to the plan's failure and the stalemate of World War I trench warfare. So, did over-reliance on a trusted technology help lead the Germans to a fatal mistake, to a false confidence that had tipped them toward military action rather than negotiation at that critical time just before World War I? Might they have hesitated if they had not regarded the railroads as highly as they did, thereby avoiding the initiation of hostilities that, along with millions of human fatalities, would also fatally wound the naive and optimistic Enlightenment faith in progress? Would the United States have attacked Iraq if it had had a little less regard for its "shock and awe" technology and a little more caution about the nature of war in the Middle East? Are these examples of category mistakes, with an unquestionably reliable Level I technology seen as therefore effective in a Level II or even a Level III environment? Certainly neither World War I nor the Iraq conflict was "caused" by technology in any direct sense, but confusion about technological complexity may have made them more probable.

It was not only in the military sense that railroads changed the course of empire. They also fundamentally altered economic and power structures, and, more subtly, cultural authority. In the United States, for example, railroads—especially the completion of the transcontinental railroad—helped validate the continental scale of the American state, and restructured the economy from local or at best regional business concentrations to trusts and monopolies by creating the potential for national-scale markets. On the global scale, railroads enabled the connection of hinterlands with ports that were themselves changing with the growth of steamship capability. This new connectivity played an important role in unifying the world economy in a way that had never before been possible, leading to the wave of globalization that characterized the late nineteenth century, and enabling economic development of continental interiors not directly served by rivers and canals.[15] With the railroad, economic power passed to industrial firms from agriculture; more subtly, so did cultural authority. The technology wave of which railroads were a major part fundamentally and radically shifted the dominant American worldview from the Edenic vision of Jeffersonian agrarianism to a technology-driven New Jerusalem.[16] But any such shift is complex and usually partial, so it is not surprising that this cultural schism replays itself even today, with the sustainability and environmental discourses leaning toward the Edenic and the industrial, commercial, and technological communities inclined toward New Jerusalem.

Perhaps the most stirring contemporary expression of the transcendence of the American technological vision can be found in Walt Whitman's 1868 poem "Passage to India." Listen not just for the exhortatory and flamboyant celebration of the technology itself, but also for the framework of meaning within which Whitman embeds the technology:

Singing my days,
Singing the great achievements of the present,
Singing the strong light works of engineers,
Our modern wonders (the antique ponderous Seven outvied,)
In the Old World the east the Suez Canal,
The New by its might railroad spann'd . . .
I see over my own continent the Pacific railroad
surmounting every barrier,
I see continual trains of cars winding along the Platte carrying
freight and passengers,
I hear the locomotives rushing and roaring, and the
shrill steam-whistle,
I hear the echoes reverberate through the grandest scenery in the
world . . .
After the seas are all cross'd, (as they seem already cross'd)
After the great captains and engineers have accomplish'd their work,
After the noble inventors, after the scientists,
the chemists, the geologist, ethnologist,
Finally shall come the poet worthy of that name,
The true son of God shall come singing his songs.
Then not your deeds only O voyagers, O scientists and inventors,
shall be justified, ...
This whole earth, this cold, impassive, voiceless earth, shall be com-
pletely justified, ...
Nature and Man shall be disjoin'd and diffused no more,
The true son of God shall absolutely fuse them. . . .

"The true son of God shall absolutely fuse them"—thus comes
the unity of God, human, and Nature, the Second Coming, in
the form of New Jerusalem, to the New World—and it comes
on rails of steel. This is not technology as economic value, or as
guarantor of national security; this is technology as salvation,
even as seen now by some transhumanists and feared (as dam-
nation) by those who reject the transhuman vision. Looked at
in another light, this is truly technology as Vishnu, "destroyer
of worlds," for the world that existed before rails, with its small
local businesses, parochial cultures, charming fragmentation of
time, small-scale capitalism, and quaint worldview of Edenic
pastoralism—that world was destroyed as surely and as effectively as ever Vishnu wrought, or nuclear winter threatened.[17]

And just like the dinosaurs, those who were there were unaware of what was truly bearing down on them, and could not have imagined the world that came after. Not unlike us.

This constellation of social, economic, cultural, ethical, theological, institutional, and policy patterns associated with a core technology is by no means unique to railroads. Indeed, railroads represent only one of what economic historians call "long waves" of innovation, with accompanying cultural, institutional, and economic changes, developing around fundamental technologies. Each core technology supports "technology clusters." Though definitions and dates are somewhat loose, major waves stand out, powered by technological clusters: railroads and steam technology from about 1840 to 1890, steel, heavy engineering, and electricity from about 1890 to 1930, automobiles, petroleum, and aircraft from about 1930 to 1990, and information and communication technology, with its computerization of the economy, from about 1990 onward (although, as we will discuss, this may be just the tip of the iceberg). And with each wave of innovation came disturbing and unpredictable institutional, organizational, economic, cultural, and political changes. Specialized professional managerial systems and associated industrial efficiency techniques ("Taylorism") characterized the heavy-industry cluster; the automotive cluster could not have occurred without a petroleum industry and a mass consumer credit system; a far more networked, flexible industrial and financial structure began to evolve during the information cluster, and so on.

The railroad story makes several general principles of technological evolution crystal clear. First, because technological systems can and often do destabilize existing institutions and power relationships, they will be opposed by many who see their place in the world and their worldviews under siege and who quite rationally seek to resist. Second, projecting the effects of technology systems before they are adopted is not just hard but, in view of the complexity of the systems, probably impossible. Little of what occurs at the frontiers of rapidly evolving technological systems is planned in advance, especially insofar as technological systems are continually co-evolving with one another and with underlying social and cultural patterns. For example, the time structure that we moderns take for granted was not the time structure of pre-railroad American or Prussian agrarian society, nor was it a result of policy deliberation and planning; rather, it was an emergent product of Level II technological evolution as society, and institutions, adjusted to the demands created by a growing and increasingly interconnected network of railroads. To talk about transhumanism without understanding the systemic transformative effect of technological change at this level is whistling in the dark. (Less metaphorically, it is to grossly overestimate how much we can know and understand about the world we live in, and how *it* is reconstructing *us*, when, frankly, most of us don't have a clue about what is going on.)

Railroads simultaneously destroyed and made worlds, and they are but one example of the ongoing human project of "creative destruction."[18] Technological change, as the example of the railroads suggests, is always potent; today, however, we have not just one or two enabling technologies undergoing rapid evolution, but five: nanotechnology, biotechnology, robotics, information and communication technology (ICT), and applied cognitive science.

Nanotechnology extends human will and design to the atomic level. Biotechnology, in the words of environmental historian J. R. McNeill, makes us "what most cultures long imagined us to be: lords of the biosphere."[19] And ICT gives us the ability to create virtual worlds at will. It facilitates a migration of functionality to information rather than physical structures; also, and critically, enhanced ICT catalyzes complexity. Money, for example, used to be coins and paper bills, themselves mere symbols of value, but now even that physical premise is gone. Money is electrons somewhere in cyberspace, and financial instruments have become so mathematical that no one can figure out anymore which shell the risk is hidden under. In 2007, the shell turned out to be the sub-prime market for mortgages, but who knew that beforehand? Not even the financial experts, it turns out. But could we have a modern economy if we had to lug metal coins everywhere? No. Even paper money would be inadequate. Money, as structured information and pure symbol in cyberspace, supports a level of economic activity that would otherwise be unattainable. Whether this is "good," and whether (and under what conditions) such a system is stable, are the sorts of questions one raises only when one recognizes a Level III system at work. Or take social networking, both in the real world and in the virtual world: As of 2010, Facebook was 6 years old; Second Life, a virtual-reality playground, 7 years; Twitter, 3 years. Yet these services have been widely adopted around the world, with consequences for written language, for cognition (e.g., multitasking), for emergent cultural and political behaviors, and for many other things that at this early point can only be guessed at.

Meaning has always been contingent on belief. Facts, in the famous formulation of William James, are simply beliefs that work, but the accelerating and unpredictable evolution of ICT creates a context for continually shifting, never-stabilizing meanings and beliefs. Of course such shifts have always been with us; the challenge now is that they are occurring in time cycles that are decoupled from our institutional and psychological ability to understand them and adjust accordingly (concepts of time, yet again, are under assault). Meanwhile cognitive science delivers its own unsettling discoveries: that "free will" as we usually think of it does not exist (data increasingly show that by the time something is in our conscious brain, our unconscious has already decided what it's going to do), that ethical frameworks are linked to particular areas of the brain (utilitarians, for example, seem to rely more on areas of the brain associated with working memory and reason, non-utilitarians more on parts of the brain associated with emotion). Wearable devices can now capture brain waves and thus enable telepathic control of avatars in synthetic reality, while Carnegie Mellon University researchers have announced a computer model that can predict what word you are thinking of. Most of these technologies and almost all of this research are primitive and are best viewed as provocative suggestions that may well lead nowhere, but their radical implications for future intervention in the domain we roughly designate as "human" are plain. And we emphasize that the domain of control they promise will remain limited to the Level I shop floor, even as their influence will pervade the increasingly boundaryless domain of the Level III Earth system.

Of course, the most radical prediction for most people is probably that of "functional human immortality" within 50 years, either as a result of continuing advances in biotechnology or as ICT and computational power make it possible to download human consciousness into information networks. And even if predictions of virtual immortality are viewed by most experts as highly unlikely (or just inane), many researchers now believe that substantial extensions of average life spans, with a high level of health, are achievable in the next few decades. For example, *IEEE Spectrum,* a mainstream technical journal, ran a series of articles in 2004 on engineering and aging which concluded that using "engineered negligible senescence" to control aging will allow average life expectancies of well over 100 years within a few decades.[20] The pervasive implications of such change, especially if rapidly achieved, are difficult to overstate: the structures of work, of families, of economics, of social status, of electoral politics, and of reproduction all are up for grabs. The idea of sustainability, especially regarding material and energy flows, would have to be entirely rethought. And this is just one element of a wildly multifaceted wave of technological change. If you can be kept healthy for well over 100 years, and biomedical technology continues to advance at a geometric rate, biological "virtual immortality" cannot be ruled out. But that's a Level I framing of a Level III problem, even at the individual level. What about boredom? What about attitudes toward personal risk—will people do anything to avoid losing virtual immortality, or on the contrary will the value of life seem diminished? What about emotions keyed to the passage of time? What about memory, which is a source of our individuality but which eventually would bog down the creativity of an immortal unless it were flushed? (Recall the pathetic immortal Struldbruggs of *Gulliver's Travels*, "opinionative, peevish, covetous, morose, vain talkative . . . dead to all natural Affection, which never descended below their Grand-children" and divorced automatically at age 80 since "those who are condemned without any Fault of their own to a perpetual Continuance in the World, should not have their Misery doubled by the Load of a Wife.") But if you do a memory flush have you "killed" an individual or "refreshed" an individual? Lurking under that superficial question is, of course, a profound one: Can selfhood exist if life is extended too far? We don't pretend to know the answer, or even to know if we got the question right, but we do know this isn't Level I turf.

We hope, then, that the concept of transhumanism, and the yea-or-nay debate surrounding it, can now be understood as desperately impoverished, and as, in many ways, just beside the point. Humans are not bundles of traits, nor are they isolated Cartesian individuals, nor are they merely hubs in social networks; they are all these things—but much more. The point is not that any particular technology may affect a particular human; the point is that we cannot understand what humans are unless we also understand the meanings of the technological systems that we make, and which in turn re-make us. For as individuals, as members of communities and larger societies, and as members of the dominant species of this planet, the state of technological play is bound up with what it means for us to be human. This was true 800 years ago, when composite bows and metal stirrups created a seamless war machine of such potency that it radically changed the structure of conflict, class, and economics. And it is true today as humans continue to dissolve the boundaries of self, machine, and nature. Familiar ways of knowing ourselves and our place, whether as part of the Great Chain of Being or through eighteenth-century political philosophy, simply do not provide an avenue for developing the wisdom and the courage we need to address what it means to be human in a world facing the revolutions in understanding and in function that flow turbulently from our irrepressible will to innovate.

Technological change, then, does not consist of isolated events marked as notches of accomplishment on the belt of progress (or decline). Rather, it occurs as movements toward new, locally stable, Earth-systems states. These states integrate natural, environmental, cultural, theological, institutional, financial, managerial, technological, and psychological dimensions; they even construct our sense of time, space, and the real. Technologies do not define these integrated Earth-systems states, except by convenience (the "stone age," the "bronze age," the "machine age"), but technological evolution can destabilize existing clusters and create conditions leading to the evolution of new ones.

Technology has always been an important, if not the dominant, means by which humans have expressed their drive to dominate. This is not just an academic observation. Cultures that develop technologies and, importantly, create frameworks within which these technologies build upon themselves and so accelerate their own evolution, thereby gain an enhanced capacity to translate volition into reliable action, and in turn achieve advantage and power over competitors. Because technologies create such powerful comparative advantages for innovating cultures, especially in a globalized society, cultures that eschew technology will, all things equal, eventually be marginalized in the process of defining the future of the

Anthropocene. And although it is impossible to imagine a plausible scenario in which technological change comes to a halt, save through a (perhaps technology-induced) catastrophe, the larger questions remains: How can human intentionality and rationality—those paragons of the Enlightenment project—be meaningfully expressed when accelerating technological evolution and complexity make a mockery of conventional notions of comprehensibility?

5

Individuality and Incomprehensibility

Those who favor transhumanism speak the language of individual choice and freedom from institutional authoritarianism; those who challenge it speak the language of human dignity and human nature as embodied in the individual. And so the transhuman dialogs center, almost obsessively, on the individual and on personal traits, as if that is the scale at which the implications of transhumanism will emerge. The hold of the Cartesian myth of the individual on our imaginations remains all-powerful: Individuals decide, individuals act, individuals make ethical choices.

We beg to differ. Consider a key assumption of the transhumanist approach: that enhanced human intelligence at the individual level adds up to a more general social benefit, so if there are more smart people then society will be smarter. The hopes and claims of transhumanism depend on this sort of arithmetic. Today, research on neuropharmaceuticals, magnetic stimulation, genetic modifications, prenatal dietary interventions, and computer-brain interfaces aims at providing direct, Level I intelligence-boosting. Yet the implied equivalence between "more intelligent" and "better" is less obvious than it might seem. On the one hand, intelligence is a complex notion, incorporating not only a person's attributes but also a society's values. But even if we allow that there will be a variety of ways to enhance a variety of cognitive capacities—concentration, memory, verbal and math skills, and even creativity—that may be components of some composite notion of intelligence in a particular human, it seems not at all necessary or inevitable that the consequence of more individuals with more intelligence will be improved humanness or humanity.

When it comes to the most difficult problems facing humanity, the main obstacle to progress really does not seem to be a lack of intelligent people. The most important problems, and those most characteristic of the irreducible dilemmas of humanness, are not amenable to radically improved solutions arrived at through rational analysis by individuals or small groups. In particular, enhanced intelligence cannot tame two essential realities of the human condition: conflict over values and uncertainty about the future.

Let's begin with the values problem. Intelligent, well-meaning people may—and commonly do—have incommensurable values, preferences, and worldviews. No optimization function exists for their diverse beliefs. In the trade-off between justice and mercy, for example, you may prefer more mercy and I may prefer more justice. In the context of terrorism, what is the appropriate trade-off between freedom and security? In the context of reproductive freedom, what is the point at which a developing embryo acquires the rights of a human being? There are no right answers. Even matters that, in our view, had been entirely settled in American society—for example, absolute proscriptions on torture—ended up re-emerging in debates over value trade-offs after the September 11 attacks, with apparently intelligent people taking committed views that we find incomprehensible and even offensive. So, to the extent that challenges to human well-being are related to disagreements about the balancing of competing values—and such conflict has bedeviled humanity through recorded history—there is no convincing reason to connect the enhancement of individual intelligence (or components of intelligence) to the achievement of more harmony in the world.

Others are more sanguine. In the spring of 2001, one of us participated in a planning workshop at the U.S. National Science Foundation for a new research program on human performance enhancement. The other participants in this small group were drawn from places such as IBM and Hewlett-Packard, from the Lawrence Livermore and Sandia National Laboratories, from the Office of Naval Research, and from the National Institutes of Health. At one point the discussion turned from machine-brain interfaces, where computers can be hooked up to human brains to augment cognitive function, to brain-brain interfaces, where the idea was that people would someday be able to communicate directly, without having to depend on the imprecisions of language. The group accepted without any discussion that such a technological capability would create a sort of two-way facilitated telepathy that would eliminate miscommunication between people and help to usher in a new era of peaceful co-existence based on mutual understanding. These were serious people in this room—real scientists and engineers, not science fiction writers (science fiction writers would never have suggested something so silly), people with greatly enhanced intelligence relative to the norm.

You can see how dumb this is, right—the idea that if only people could see with perfect clarity what is in other people's brains, then they would understand one another and get along better? But what if people held conflicting values or interests or ways of understanding how the world worked? Would knowing what was going on in the heads of people with whom one disagreed be a path to harmony, or to conflict? Imagine two seasoned diplomats, say one an Israeli and one a Palestinian, engaged in tense negotiations, and with direct access to the other's thoughts. Does anyone seriously think that peace and mutual understanding would somehow automatically arise? Isn't the ability to hide thoughts and emotions a crucial skill in diplomacy?[1] The cognoscenti at the workshop were somehow missing the fact that what they saw as imperfection in language and communication—something to be corrected through enhancement—also provides the subtlety, malleability, and ambiguity that in fact may keep people talking to one another rather than killing one another—attributes around which complex social institutions (e.g., political and legal systems) have been evolving for millennia. Raising these points didn't change the brilliant minds of the men (yep, all men) at the workshop. About a year later, the group—again, under the sponsorship of the National Science Foundation—issued a report that included a utopian vision of performance-enhancement technologies that "could achieve a golden age that would be a turning point for human productivity and quality of life" such that "the twenty-first century could end in world peace, universal prosperity, and evolution to a higher level of compassion and accomplishment."[2] As we write, it seems almost too poignantly apparent to note that, a decade into the new century, we're off to a pretty lousy start.

The second, quite related reality of the human condition that stubbornly resists domestication through greater intelligence is uncertainty about the future. No one knows how to intervene in complex social, human, built, and natural systems to reliably yield particular desired results over the medium or the long term. How did all our advanced economic modeling and theoretical capacity help us to avoid the 2008–09 global economic melt-down? In fact, overconfidence in such models and theories helped create the problem. On a wide range of subjects—ecosystems management, weapons non-proliferation, organizational management, immigration policy, improving the conditions of our inner cities—hundreds of thousands of academic publications have certainly added in some sense to our intelligence, but without adding much to our capacity to act with consistent or increasing effectiveness. That there is a remarkable absence of increasingly effective practice in these broad areas of human affairs, despite all the effort aimed at better understanding, is not a statement about the limits of our intelligence but rather a statement about the limits of the type of intelligibility that can reliably guide action when the future is uncertain and values conflict. Cause-and-effect chains simply do not extend very far from the present. Pick any individualistic transhumanist formulation you want and this underlying uncertainty will not change. Transhumanism, it turns out, cannot be the solution to the techno-human condition; it is merely the latest manifestation, another category mistake trying to convince us that, by playing with a subsystem, we can change the larger system, and its emergent behavior, in ways that are *a priori* predictable and desirable. No can do. If you want a new measure of rationality in this world, one that suits the complexity we are creating, you will need new concepts, new tools, new arrangements, and perhaps even new gods to replace those old ones like individuality, rationality, predictability, and the like. But that's for the last chapter.

When people convince themselves that they are smart enough to escape from the systemic limits of incommensurable values and unpredictable futures, watch out. As the incomparably wise Albus Dumbledore has noted: "I make mistakes like the next man. In fact, being—forgive me—rather cleverer than most men, my mistakes tend to be correspondingly huger."[3] Consider, by way of illustration, the careful and persistent deliberations of a small group of influential people whose intelligence had been supremely enhanced relative to the norm via privileged education and rarified social networks: the Neoconservatives who convened around the Project for the New American Century in the late 1990s, many of whom then rose to high political posts in the administration of President George W. Bush. The insular deliberations of this group of extremely intelligent men gave rise to the theory that justified the war in Iraq—a theory of democratic nation building that looked good for a few months but then turned out to be incapable of encompassing the widening gyre of consequences of the war—consequences that in turn helped to undermine the power of the

Neocons (and, through debt and leaching of moral standards, that of the United States). If this example strikes you as partisan, you might prefer to consider the super-intelligent Kremlinites who thought it was a good idea for the Soviet Union to invade Afghanistan but instead helped to hasten the demise of their own empire. Or, forgetting about war, consider the two Nobel-prize-winning economists who helped to found the risk-free hedge fund Long-Term Capital Management but were unable to anticipate the downturns in the East Asian and then Russian economies that led to the fund's collapse in 1998 after it incurred a loss of $4 billion, or the brilliant systems dynamics theoretical structure developed for complicated technology design environments by Jay Forrester at the MIT Sloan School of Management that was then applied to urban systems, where it has generally failed; or the distributed cognitive network of bankers and corporate investors that gave mortgages to millions of people who might not be able to pay them back because, after all, housing prices could never go down! All these cases involved already-enhanced people—the best we had—and they tell us unequivocally that individual enhancement is not social rationality. When it comes to figuring out how to manage and govern our growing capacity to enhance our individual selves, we must face a level of institutional and environmental complexity that the transhumanism dialog, in its naiveté and superficiality, has yet to begin to comprehend. Neither has (nearly) anyone else, we hasten to add—which is precisely why the transhumanist project has proved to be such a useful exploratory tool for us; its weaknesses are apparent, but they are mere proxies for the weaknesses of our current intellectual and cultural frameworks. These failings cannot be overcome by making us all smarter, but only be reinventing the underlying framework: the Enlightenment commitment to rational action by individuals living in a comprehensible world.

When the system is complex, and when values conflict about what is to be done (the two go together, of course)—when, in other words, we are at Level III—muddling through is often the best we can do. Progress, when it occurs, comes through trial and error, through learning what works in particular situations, through incremental change that incorporates such learning, and through the difficult process of political compromise that allows people to take a next step. Complex, value-laden problems such as immigration, environmental degradation, health-care-system dysfunction, the global drug trade, and conflict in the Middle East don't get solved; at best they get managed, and at worst we lurch from crisis to crisis. What the political theorist Charles Lindblom called the "intelligence of democracy" is not a summation of IQs that allows "smarter" societies to arrive at the right solution to a complex problem fraught with value conflict and complexity, but a mélange of diverse world-views and value structures that keeps democratic societies—some of the time—from doing anything too stupid. Obviously this does not mean that a nation of chimps would do as well as a nation of geniuses, but it does mean that a nation of geniuses needn't do any better than the standard normal distribution of cognitive capacities that we have at our disposal right now. Intelligence must co-evolve with, and emerge from, experience. And because the techno-human context is always changing, the lessons of yesterday's experience are not easily transferred and applied to today's problems. When the hubris of intelligence gets out in front of what is learned from direct, contextual experience, the results are often disastrous.

We cannot enhance ourselves out of this situation any more than we can enhance ourselves out of the passage of time or the increase of entropy. The challenge is to our political and social institutions, not to our individual intellects. Because transhumanism cannot escape from the grip of individuality and the idea that society is a simple sum of individual attributes, it cannot grasp this fundamental point. Thus, for example, the transhumanist James Hughes proposes an automatic, additive connection between human enhancement and democracy, which leads him to argue that "increasing human intelligence will encourage liberty [and] democracy" and that "the more intelligent the citizen, the more capable they will be at assessing their own interests, understanding the political process and effectively organizing."[4] Hughes treats "intelligence" as some single attribute whose enhancement will inevitably yield a particular outcome. What about people of great intelligence who have been leaders of some of history's most authoritarian movements—perhaps for exactly the reasons Hughes states? Even if some sort of enhanced intelligence did allow people to better assess "their own interests" (which seems like a strange claim to make), it isn't as if "one's own interests" is a simple thing that, once recognized in a particular way, will lead people to accept the legitimacy of advancing those interests democratically. Why wouldn't the opposite be the case—that people would, as they do now, seek whatever means they could find to advance their interests—in which case democracy and rule of law would remain the only antidote to the efforts of particular groups of enhanced people to pursue their interests at the expense of others? Nobody knows, or can know, what the best distribution of intelligence in its many forms might be for a healthy democracy. As the conservative intellectual William F. Buckley Jr. once famously asserted, it is better to be governed by the first 2,000 names in the Boston phone book than by 2,000 members of the Harvard University faculty.[5]

The incommensurability of human values and value systems, and the real-world complexity that makes it so difficult to know how actions in the present will connect to consequences in the future, are direct and fatal challenges to the belief that technological enhancement of human cognitive capacities will chart some new and improved path toward better humanness and humanity. Humans do not live lives unconnected to other humans, and the outcomes of human enhancement will depend on the world into which enhanced traits are inserted, not, as our enhanced soldiers in Iraq tell us, on the enhancements themselves.

If we were to imagine a better world, it would not simply be one in which memory or concentration in some people were improved, but one in which humans and humanness were better. It would, for many, be a world with more justice, more equality, more peace, more freedom, more tolerance, more friendship, more beauty, more opportunity. Such conditions, and the social and political changes that could encourage them, cannot be internalized in the technologies of human enhancement, or in any technologies of which we are aware[6]; even less can they be designed to emerge from the aggregate effects of enhanced individual traits in many humans. These are Level III goals, and no amount of hand-waving will make them achievable through direct application of Level I technologies. Transhumanism and the technological program for human enhancement turn out to be the mirror of, not the cure for, the techno-human condition. Put another way, if we define the goals of a particular enhancement technology in an appropriately limited manner—say, enhancing memory in older people to help them function at a high quality of life—it may be possible to build a shop-floor technology, such as a neuropharmaceutical, that meets that goal and, in that context, represents progress. But if we define the goals as the transhumanist dialog has tended to do—i.e., to build a better and more equitable world in which democracy flourishes because people can remember more things and can even intuitively understand one another's thoughts—we've blown it. We have confused the shop floor with the complex adaptive system, Level I technology with Level III. We are back in category-mistake country.

Simply consider what is being enhanced. With neuropharmaceuticals, we can increasingly enhance particular aspects of an individual brain, a Level I coupling of technology and goal. But even as these technologies advance, cognition itself increasingly looks like an integrated activity across technological networks. This is not new: Edwin Hutchins, in his excellent book *Cognition in the Wild* (1995), analyzes pre-GPS naval navigation processes to make the point that technologies, in this case naval charts and tide and current tables, and artifacts such as compasses, not only redistribute cognitive workloads across time (charts don't have to be remapped for every fix) but also dramatically simplify the cognition that must take place in real time: "Rather than *amplify* the cognitive abilities of the task performers or act as intelligent agents in interaction with them, these tools *transform* the task the person has to do by representing it in a domain where the answer or the path to the solution is apparent." (p. 155) Thus, "humans create their cognitive powers by creating the [cultural and technological] environments in which they exercise those powers." (p. 169) We can think of this, if we want, as making cognition a combination of congealed and real-time elements. Students writing papers by coupling to the essentially unlimited interactive and continually evolving and growing memory of Google are thus combining congealed cognition (the hardware and software that give them access to Google) with real-time cognition (in the combined form of their internal cognition and the real-time cognition provided by Google software and hardware platforms in responding to their queries). This is clearly Level II cognition, and it is far more complex than simple Level I pharmaceutical enhancement. It is also Level III, because we have very little idea what the cultural, institutional, social, and psychological effects of these dramatic increases in cognitive networks will actually lead to—it is, after all, not just Google, but also social networking, augmented reality, augmented cognition (such as self-operating cars), and a myriad of other technologies that are integrating at this point in our history.

This confusion of levels will not be an obstacle to the proliferation of human-enhancement technologies. One can hardly doubt that many people, perhaps most, will avail themselves of all the enhancements they can afford and can stomach if they believe they will individually benefit in some way. However, we can best understand this process not as the noble pursuit of better humanness in some larger sense, but as the continuing human desire to be stronger, smarter, better, with perkier breasts and a flatter stomach, than one's peers, advanced as the usual brand of consumerism, advertised as self-improvement, embraced by hope, enforced by the fear of falling behind, and indefinitely sustained by the certainty that tomorrow's enhancements will soon come to feel boringly normal—the civil union of Narcissus and Sisyphus (except where prohibited by law). Meanwhile, the underlying science and technology necessary to feed the consumption compulsion will be driven by the ongoing quest by technologically advanced states for military and economic advantage.

So let us dispose here of a tempting analogy between transhumanist claims (rooted as they are in defense of individual liberty) and Adam Smith's famous "invisible hand." Just as innovation and productivity are optimized as individuals, driven by self-interest, strive for economic benefit, won't society be made better by individuals striving after various enhancements? The analogy would be false. The "invisible hand" is just a recognition of an information-processing mechanism; it integrates decisions made on the basis of individual preference into supply and demand curves that describe the efficient distribution of scarce resources. Unlike central planning, the "invisible hand" is effective with complex adaptive systems (i.e., the economy), in part because it decentralizes decision making across active agents and in part because the only outcome it aims to deliver is efficient allocation of information, prices, and stuff. But—and this is the critical part—the market enables efficient use of resources, not a fair, or even stable, distribution of resources across society—it does not guarantee that the world will get better, and Adam Smith well understood this fact. The early history of capitalism was one of violence and poverty alongside great wealth, huge social dislocation, and oligarchic monopolies, as documented in the novels of Dickens, Dreiser, and many others. A more socially stable and, yes, advanced (if still highly imperfect) capitalism evolved only when dampers on rampant market freedom, such as unions, anti-monopolistic policies, and unemployment insurance, were adopted—almost always after bruising political battles and occasional paroxysms of civil unrest. There is no invisible hand to guide society toward more justice and tolerance, with or without radical enhancement of human capabilities.

But now there is another possible complication to this story. The invisible hand of the economic market works because we can assume that humans generally behave as they always have: selfishly. Might human-enhancement technologies threaten even that fundamental assumption? (Put another way, could we design "selfishness" out of the human?) We doubt it, yet the larger point is this: Even if it seems as if we are simply modifying the constitution of humanness at the individual level, the systems-level effects of tens of millions of such modifications may plausibly begin to manifest in system-wide changes in human values and behaviors that cannot possibly be predicted. Markets are an information-processing mechanism built on, and assuming, stable legal, cultural, and institutional foundations, and a stable idea of what humans are; enhancement renders those foundations contingent and unpredictably changing. Markets assume a particular context; enhancement transforms it.

Like the network of vaccination that provides group immunity, or the network of communications that makes your cell phone more than just an over-engineered paperweight, enhancement technologies do not live quietly at one particular level. Plastic surgery provides individual enhancement, but it also drives social norms about what is considered attractive at what age. In order to remain competitive, professionals competing for limited spots at desirable law firms will be driven to cognitively enhance themselves as their peers do. Soldiers in combat are enhanced to increase their effectiveness and endurance, and the efficiency with which they can be coupled into battlefield technologies, perhaps even at a distance, becomes an important determinant of their lethality and their survival. India, China and Brazil may view genetically modified crops as important for feeding their populations and, concomitantly, climbing the league tables in cultural authority. An individual decision not to take Botox, not to enhance one's mental acuity, not to kill one's enemies more effectively, or not to consume transgenic foods can end up, through the multiplier of millions of such decisions, transforming a person's social and economic life, or a country's military and economic competitiveness, in ways that make a mockery of notions of individual choice and autonomy.

How might this happen? Let us pose a scenario that is directly contrary to the idea that rapid technological change and individual enhancement will augment democracy. Suppose that rapid technological change and enhancement, combined with a concomitant failure to evolve ourselves and our institutions so as to better comprehend and manage our unflagging ingenuity, ends up threatening the sustainability of democracy. What if—as uncomfortable as this scenario may be for those who (like us) view various approximations of "government of the people" as essential ingredients of a good society—we are beginning to experience the end of the great Enlightenment project of radical democratic power? How might this happen?

To begin with, the shared experience of radical technological transformation—of life in Level III—is confusing and challenging, and provokes among some groups a turn toward fundamentalist certainties that can offer the balm of social and psychological stability amid spiraling technological complexity. Fundamentalism is on the rise in virtually all major religions, as well as in certain belief systems—e.g., environmentalism, neoconservatism—that for many people, especially in secular societies, offer certainties that serve essentially theological purposes. We view this not as random opposition to modernity, but as a social phenomenon reflecting a political reality: as the reach of technological transformation expands, increasing numbers of people in every society are disenfranchised from having a say in how that transformation occurs, what it means, and who gets to benefit from it. They are incapable of keeping pace with continuing technological change, unable to integrate into the information webs that increasingly define human cognition, and aghast at the changes in lifestyle, income distribution, relative power relationships, and sexual and family roles and structures. Fundamental values are being rendered contingent by technological progress, and for many people this is not a good feeling. Thus, even as technological transformation is a central component of cultural (not to mention economic and military) dominance, so does it provoke opposition to itself. As we saw with the September 11 attacks, advanced technologies themselves become tools of opposition to cultural dominance built on those same tools; they become tools to assert identity that rejects that dominance.

Earlier in this book we traced ideological framings of the transhumanist debates to their origins in religious traditions. These origins are, of course, camouflaged because the debates are so often carried out in the language of science and technology—the language of power in the modern world. The evolution of human technological competency in nanotechnology, biotechnology, and other emerging capabilities, potentially subjects the entire material world, including the biological world (which of course also includes the human itself, both physically and cognitively), to human design. This nascent capability challenges cultural and religious assumptions about appropriate boundaries between the sacred and the human, and conspicuously emerges as a theme among those for whom "Nature" has become the repository of the Sacred, a continuing reflection of the Romantic project to protect God from science by shifting the Sacred to the wilderness.[7] The devolution of God into Nature is an important foundational belief for many environmentalists, ranging from English Royals who perceive biotechnology as blasphemous

because it is "playing God" to environmental writers such as
Bill McKibben, who first places God in "nature" and then be-
moans the human impacts on the latter: "Wild nature, then, has
been a way to recognize God [of the Christian tradition] and
to talk about who He is. How could it be otherwise? What else
is, or was, beyond human reach? In what other sphere could a
deity operate freely?"[8]

But might the rapid evolution of technologies that have the
capacity to render widely shared notions of "the human" contingent in the face of Level III consequences (consequences that
play out so rapidly that everyone can see them happening) lead
toward an increasingly final rejection of the roles assigned to
deity, human, and beast in many religious traditions? Rebalancing theological interpretation and scientific advance has
long been a critical discourse, and authorities dating back to
Saint Augustine have offered solutions. (Augustine, for example, noted the necessarily unitary truth of science and theology
under an omniscient God, which led to his conclusion that science and theology, far from being adversarial, were necessarily
complementary—a position later shared by Copernicus, Kepler,
and Newton.) Consider, for example, Pope John Paul II's comments in the 1998 encyclical *Fides et Ratio* (introduction and
paragraphs 34, 43, 48):

Faith and reason are like two wings on which the human spirit rises
to the contemplation of truth. . . . The two modes of knowledge lead
to truth in all its fullness. The unity of truth is a fundamental premise of human reasoning, as the principle of non-contradiction makes
clear. . . . Both the light of reason and the light of faith come from
God. . . . Hence there can be no contradiction between them. . . . It is
an illusion to think that faith, tied to weak reasoning, might be more
penetrating; on the contrary, faith then runs the grave risk of withering into myth or superstition.

We wonder if rapid change at the Earth-systems level of technology will lead—is, in fact, now leading—to a more radical
redistribution of ethical and moral responsibility among categories that are of great concern to many people: the sacred,

the profane, the humanist, the realm of God alone. Genomics, *in vitro* fertilization, technologies that can keep bodies living long after mind ceases to function (and perhaps vice versa), and other changes to traditional concepts of what "human" and, indeed, "life" are, will not only be controversial, but will also shift increasing responsibility from God and the sacred domain to humans and their secular institutions. The shift is already happening, of course: Who, today, more often decides when death occurs—a priest (a servant of a church), or a doctor (a servant of a health-maintenance organization)?

Different communities will perceive change, and alter doctrines, at different rates. Secular institutions are likely to be more nimble than religious ones, whose authority is more clearly tied to existing roles and categories. And, as more conservative communities begin to understand that new patterns of integration of the traditional "human" and technology systems are already here, and that technological evolution generally is strongly correlated with cultural and economic power, their reaction will intensify, perhaps dramatically. This is nothing new: the original Enlightenment was characterized by just such a redistribution of responsibility, and, at a deeper level, by a fundamental shift in the cultural source of knowledge, from given authority to knowledge based on observation and experiment. But the pace and the scale of change may be unprecedented.

And so, in adopting the goals of religion—dominion, perfectibility, transcendence, and so on—transhumanists and advocates of human enhancement are in one sense only continuing to extend the gradual convergence of faith and inquiry that arose from the Enlightenment's commitment to applying rational means to the improvement of human affairs. Yet one cannot help but perceive emotional emptiness in a worldview that sees, for example, transcendence manifested in the downloading of the contents of one's brain into networked computers and that somehow misses the point that the ends of religion are important to the worldly affairs of humans because of the means they can cultivate for pursuing a moral life, not because of the rewards they deliver in the hereafter (although, to be fair, much contemporary evangelical religious practice seems to display a similar bias). On the other side, the transhumanist critique that relies on a socio-theological privileging of the current cultural construct of "human" and on the notion that a visceral "yuck factor" provides sufficient reason to challenge adoption of transhumanist technologies[9]—a sort of updated "just say no" policy—is not only a partly cultural phenomenon (Americans may not like humanoid robots, but the Japanese do) but also apparently oblivious to the promiscuity of market capitalism. The diffusion of enhancements through society, whether it's steroids, plastic surgery, artificial joints, or antidepressant medications, can be impressively widespread, often with little regard for legality, safety, or even effectiveness. That opponents may recognize this human tendency to sin is demonstrated by their fitful efforts to legally ban enhancements. Indeed it was always thus: Stop me before I enhance again. And stop everyone else as well.

Opposition to technology is, of course, an honorable historical tradition, and technologies from the stirrup to the printing press to the railroad have provoked resistance while they were transforming societies to their roots. But what may be unprecedented now is the rapidity, the cultural reach, and the global scale of the cultural transformations themselves, and, in turn, the magnitude and ferocity of social response, and, on a larger scale, the implications for cultural competitiveness.

Thus does technological change create an excruciating tension. Can liberal democracies—flanked on one side by bellicose fundamentalism and on the other by countries such as China, whose commitments to rapid economic technological advance may at best render pluralistic discourse an inconvenience to be swept under the rug—steer a middle course of technological moderation and still maintain their ability to compete in the global marketplace? Or will the advantage go to less open cultures in which the elite, who benefit from technological evolution generally and from human enhancement specifically, are able to increase the effectiveness of their control over opposition? Will such an enhanced elite, skilled in navigating complex and information-dense environments and irrevocably committed to breakneck technological advance, become more and more dominant while the rest join an expanding global proletariat? Or what if China—the country that restricted families to having only one child—decides next to require all those children to take performance-enhancing drugs as part of the national approach to education and training?

Why do we suggest this scenario? For several reasons. Most important, perhaps, it further illustrates how contingent the argument that transhumanism strengthens democratic ideals and performance is. We don't know which scenarios are more probable, but we do know that any particular projection, utopian or dystopian, is likely to be wrong at this point. We must be careful not to reify our pet scenarios.

Beyond that, the scenario illustrates a basic dynamic that we think is both amusing and important: The complexity and contingency of the modern world is due in large part to the success of a traditional Enlightenment framework that helped give us democracy in the first place. Without democracy and the constrained conflict it usually entails, it is less likely that we would have seen the evolution of the technological, social, cultural, and institutional complexity that is the foundation of the transhumanist discussion, and of the complicated questions with which we now grapple. For example, one source of political complexity that is attributable in part to implementation of Enlightenment values in the social realm is the exponential growth of non-governmental organizations, interest groups, and Internet-based clusters. This complexity is indeed a tribute to technology as politically empowering, to the idea of the individual as politically meaningful, and to the idea of rights as accruing not to classes or elites but to all. But it also creates dynamics at Level II and Level III that significantly constrain individual choice and intentionality, and limit the usefulness of worldviews and admonitions based on the systemic efficacy of the individual. (That's why "Think globally, act locally" doesn't work: it is a category confusion between Level II and Level I.) The Enlightenment, in other words, has not failed; it has perhaps succeeded too well in birthing a world that now requires us to move beyond the intellectual and cultural tools it first provided us.

Though we wouldn't dare to handicap the odds that continued breakneck technological change and human enhancement threaten national systems of democratic governance, the question serves to illustrate both the non-intuitive possibilities that may lie in our future and the profound pedestrianism of the transhumanist dialog (which is locked so firmly to past verities). Few would argue the observation that large-scale technological transformation is a process far beyond the governing grasp of democratic decision processes as conventionally discussed. Though this is true of other forms of government as well, it may be that democracy, because it better translates backlash against technology into political blocking power, could be less adaptive than other cultures. (Time will tell.) Transhumanists may posit an automatic evolution toward better democracy; we want to highlight that, in the world we are continually creating, nothing is automatic or obvious.

Finally, an important additional reason for introducing this scenario is that we want to encourage the practice of building intellectual option spaces so as to provide room for thought experiments that, though undoubtedly incorrect in specifics, nonetheless offer practice in thinking about potential futures and about what our institutional and policy responses might be—to build resilience and adaptability into our culture. We need to substitute "explore with humility" for "attack with rigidity." This is similar to what militaries do with war games, because they know, in von Moltke's famous phrase, that no plan survives initial contact with the enemy. The only way to prepare is to build tactical and strategic agility in the face of uncertainty. For just those reasons, many firms use scenario methods to enhance their understanding of the complex landscapes they face and to identify options that might be useful in various circumstances. Thus, the appropriate question is not whether the scenario of rapid technological change placing democratic societies at a disadvantage relative to more autocratic competitors is right; rather, it is whether that scenario usefully helps us think about various options and actions, and their implications, in view of the unknowability of the future.

6

Complexity, Coherence, Contingency

Not everything is complex. In particular, when one is dealing with a shop-floor, Level I technology, one is dealing with a simple system.[1] By that, we mean that most of the necessary relationships among goals, means, and causality have all already been captured in a physical system that can be used with confidence that a given input will produce a desired output. If you are vaccinated against measles or tetanus, it is highly probable you will be protected from the disease for the specified time; if you get into a car, it is highly probable that you can then drive on a road to the location desired; if you take an appropriate neuropharmaceutical, it is likely that you will do better than you otherwise would on the mental task that the drug is aimed at enhancing. All these things create a sense of greater control and individual effectiveness.

We lose this simplicity with Level II, networked technologies. An airplane works, but bad weather or the failure of an air-traffic-control computer keeps it on the ground. The citizens of a developing country are vaccinated, but a lack of job opportunities and necessary infrastructure does not allow improved health to translate into economic growth. A commuter gets into her car, then sits in traffic and misses an important meeting. With Level II technologies we begin to experience complexity that is often surprising and unpredictable, but it is complexity that we can understand. With Level III technologies operating at the scale of Earth systems, however, fundamental cultural and institutional verities—what we believe to be true; our ways of knowing and making sense of the world—may be rendered contingent and even nonsensical. The rise of railroads signaled the decline of the Jeffersonian agrarian worldview. If suites of technological enhancement of human performance come on line, the individual as we now conceptualize it may change profoundly and unpredictably, rendering contingent many of the cultural and institutional structures that presuppose a particular type of individual or a particular set of virtues and beliefs about what individuals ought to do—the foundations, that is, of how we make sense of and operate in the world.

All complexity, however, is not the same. The most elementary complexity is static complexity arising from increasing numbers of hubs and links in the system of interest—more components, stakeholders, interactions among different infrastructure, and linkages among them, for example. Static complexity characterizes many Level I technologies—there is nothing simple about the number of parts and linkages in a modern jet aircraft or the number of connections and operations built into a computer chip. As the hubs and links in a system interact over time, complexity may become more dynamic, both internally and in interactions with the external environment, in new and unanticipated ways. One does not need, however, a complex static structure to give rise to a complex dynamic one (one reason why static and dynamic complexities are not the same). MIT's famous Beer Game, in which business students try to balance only four nodes—a retailer, a wholesaler, a distributor, and a factory—and almost always fail miserably, is an example of a simple system giving rise to unpredictable dynamics (many of which arise from time lags built into the very few pathways of information flow, which in turn increasingly compromise the ability to make good decisions—you can try your hand using the online version at http://beergame.mit.edu/).[2] Nonetheless, many Level I systems are able to cope with dynamic complexity; for example, not only is an airplane very complex from a static viewpoint, but it flies in a lot of different environments—differing maintenance regimes, different climates, sometimes extreme conditions—and does so reliably. Successful technologies are designed with such complexities in mind, thereby internalizing dynamic complexity, or rendering it irrelevant, to a remarkable degree.

When complexity becomes "wicked," all bets are off. "Wicked" here is a term apparently first introduced by Horst Rittel and Melvin Webber in the early 1970s to distinguish the deep differences between natural science complexity, and sociological and cultural complexity. Complexity is wicked when a system's makeup and dynamics are dominated by differing human values and by deep uncertainty not only about the future but even about knowing what is actually going on in the present. Any solution to a wicked problem should be expected to create unanticipated but equally difficult new problems—a crucial insight, in view of the brash assurance of the transhumanist dialogs. Why is wicked complexity so difficult to manage? Well, to begin with, the systems giving rise to the problem cannot be definitively described or defined. Drawing boundaries around such systems is necessarily arbitrary; any effort to define the system thus oversimplifies other aspects of it, and downplays or ignores critical couplings between what you have defined and other, external factors. Thus, for example, the current penchant for defining climate change as a primarily environmental issue, rather than as a complex and difficult social, economic, and cultural condition, is one reason why policy initiatives such as the Kyoto Protocol have failed.[3] Defining cities as discrete entities that must be managed for particular outcomes (e.g. environmental quality), rather than viewing them as emergent phenomena characteristic of our species, with myriad dimensions and domains, is another obvious example of dysfunctional (and amazingly unintentional) oversimplification. In the same vein, the naive utopianism or dystopianism of the transhumanist dialogs arises from a failure to understand that the techno-human condition embeds us irretrievably in wicked complexity. In such systems, there is nothing like an indisputable public benefit; only through ideologies or other simplifying mechanisms can one even pretend to have "the answer." Marxism, or neoconservatism, or environmentalism, or some version of Christian orthodoxy, or even science, can give you an answer. But when problems are wicked, no belief system, however rooted in analysis of facts it may be, can provide a universally accepted answer (which is why in many cases such "answers" are accompanied by coercion, usually subtle and non-violent in a democracy). And so, of course, there is no "correct" policy or resolution to a wicked problem, nor is there optimality. There is only, as we have said, muddling through. And, importantly, muddling through is not a second-best process to be dropped when appropriate optimization techniques are developed: it is the best we can do. Remember what Dumbledore said.

Problems of wicked complexity are compounded at the scale of the systems humans are now engaging, affecting, and being affected by. The world is composed of increasingly integrated human-natural-built systems that display, on regional and global scales, the interaction of decisions made in many different jurisdictions with many different, often conflicting goals in mind. Something that may be unimportant in one area, such as how much atmospheric nitrogen is being deposited on agricultural land, may be quite important elsewhere (for example, in an estuary), and one society may be seriously engaged with questions of global climate-change policy while another is simply trying to find enough food and relatively clean water to keep people alive and the economy growing. Under such circumstances, nostrums such as "Think globally, act locally" are naive and unhelpful, for the simple reason that good local decisions don't add up to good global outcomes. We are entering a new domain in which ethical and responsible behavior as judged by outcomes in the real world is an increasingly meaningless idea, at least given the simplistic version of rationality we have been employing for several hundred years.

Wicked complexity just can't be successfully managed using that favored child of the Enlightenment, applied reason. To further explore this point, let's stick with climate change for the moment. Enormous amounts of data on the climate system have already been generated, and understanding is still highly uncertain (even if certain phenomenon, such as global warming, are well documented). Scientists employ increasingly complex mathematical models, known as general circulation models (GCMs), to try to understand how the system behaves. So long as these models are regarded as quantitative scenarios, they are useful: they allow us to conduct thought experiments and to develop potential policy responses to a wide range of plausible futures. But because those who are concerned about the consequences of climate change feel a powerful need to make strong claims about the future, they often treat model results not as scenarios but as plausible futures, and climate science not as an input into more complex social deliberations but as the determinative discourse.[4] Politically, a major reason why top-down international approaches to climate change exemplified by the Kyoto Protocol have failed is that they represent the effort of Earth scientists, environmental advocates, and diplomats to take a complicated human system and manage it through the applied reason of climate-change science. Much of the scientific information that justifies the Kyoto Protocol has been provided by the Intergovernmental Panel on Climate Change, whose many products include multiple scenarios about, for example, future temperatures and carbon dioxide emissions—scenarios that are always represented as suites of smooth curves, based on different assumptions, extending for the next hundred or more years. And yet, though it is not possible to predict the specifics of technological evolution over that period, we can say with a high degree of confidence that any such curve is surely wrong, starting with its smoothness.

Imagine trying to predict out a hundred years in 1900. Anyone who may have been considered an expert at that time would have missed nuclear weapons and energy, computers, aviation, television, the Internet, bioengineering and other technologies based on the genome, and the "Green Revolution" in agriculture that wrong-footed so many ecologists who, in the 1970s, were predicting massive global famine by 2000 (not to mention so many agricultural economists who, on the basis of the Green Revolution, thought that world hunger would vanish).

Furthermore, such scenarios are self-negating: if people don't like the future that is being created, they will act to change that future. The future will absorb predictions, change its behavior, and invalidate the prediction—though in unpredictable ways, of course.

Finally, the scale and the scope of the climate system—the fact that it is interconnected to virtually everything else, including the nitrogen and hydrologic cycles, ocean circulation patterns, human economic activity, and political ideologies—means that anything humans do to try to manage "the climate problem" ramifies across many other systems, most of which are not being considered at all. From the Level III perspective, the inadequacy of current policy and institutional approaches for coping with human impacts on the climate system—from computer modeling to scientists as philosopher kings to the United Nations to our mental models, ideologies, and worldviews–is all too apparent. In the simplest version of the problem, many people still seem to believe that applied reason, as manifested in a move from Level I (solar panels, wind turbines, nuclear reactors, whatever) to Level III (manipulating the climate in desired ways) is only a matter of doing the right thing, installing the right type of solar panel, buying organic produce, making virtuous choices. Indeed, this is the only place the Enlightenment can leave us. And here, so far, is the score: Nearly 20 years into the effort to impose a global governance regime on the climate, no progress has been made toward reducing global emissions. In fact, even by that excessively narrow definition the problem is growing worse. This empirically robust result is not attributable to "bad actors" such as the United States; rather, it is a sadly unavoidable—predictable—consequence of category mistakes—of Level I and Level II thinking in a Level III world.[5]

When it comes to human-enhancement technologies, we are occupying the same space (and making the same category mistakes). In a Level I analysis, evaluation of such technologies is straightforward: if a memory-enhancing drug improves memory in a double-blind study, it is effective (probably). But if a lot of people use it, that implicates Level II behavior. The consequences may still be understandable, but they can no longer be internalized in, or predicted from, the technology itself. For example, suppose one wanted to improve the performance of soldiers as a group, or the quality of engineering in a particular society. These goals cannot be accomplished simply by enhancing the intelligence of individual soldiers or engineers. They would require changes in military tactics, technology, and strategy, in the education of engineering students, and in the salaries paid to engineering graduates. The goals, in other words, may be clear, but they are not determined by the functionality of the technology alone—and the desired institutional and organizational changes may be much more difficult to make, with consequences much more difficult to predict, even if we discount the countermeasures they may provoke. Whew! And we're still at Level II, where, at least, we can hold on to our fundamental moral and factual moorings. What then happens as wider use of some enhancement technology changes what we expect "human capability" to be, and perhaps establishes different groups and shifts competitiveness among cultures, moving us toward Level III, where the goals we started out with are no longer even clear, and where uncertainty and contingency may even undermine assumptions that would make it possible to define goals in the first place? At that point we enter the realm of wicked complexity and global implications—we are in Level III.

One of the key experiments of the twentieth century was the society-wide application of applied reason to the wicked complexity of economics through Marxism, the apotheosis of Enlightenment rationality. Wickedness won. Marxism in the Soviet Union and in China collapsed not from external conquest, or as a result of President Reagan's spending, but because, above all, the centralized economic model adopted by large Marxist societies proved incapable of managing the complexity inherent in a modern industrial economy. Even Gosplan, the State Planning Committee of the Soviet Union, could not understand, much less micromanage, the relatively stunted economy of the Marxist empire, a task that the post-Enlightenment decentralized information structure we know as the market took care of for Western opponents of state communism.[6] Our point is strengthened by the example of China, which has held on to political Marxism only by surrendering the illusion of absolute control of the economy. And, of course, economies, financial networks, and technologies have become far more complex since then—to the point where, as the Panic of 2008 showed, even the most sophisticated investors, financial institutions, and regulators completely lost track of the risk packaged in modern financial instruments such as mortgage-backed securities and credit-default swaps. We cannot centrally control the global economy; indeed, it may be impossible to centrally conceptualize it for much longer without escaping from the one thing that seems most essential to our modern identities: the belief in our ability to act on the basis of rational understanding. So, yes, perhaps the ruckus about transhumanism does signal the need for a profound change in humans and their institutions—but if it does, the need is completely different from what the transhumanists and their enemies have in mind.

The complexity we are discussing confirms an unavoidable relationship among the observer, the frame of reference, and the extraction of partial and contingent truths from underlying complex systems. Put simply, if a model, a worldview, or a system is internally coherent and intelligible, it is necessarily partial. Consider a trivial example: If we are interested in the rate of violent crime in New York City, we implicitly define the urban system by its political boundaries, because violent crime is a geographically localized event and statistics on such events are organized with reference to political boundaries. If, on the other hand, we are interested in water supply and demand in New York City, we are implicitly defining the system as including the watersheds of most of New York State, the land-use laws that have been significantly shaped by New York City's need for water, and the physical infrastructure that has been constructed and maintained, and legally regulated, to supply such water. Yet in both cases the relevant marker is "New York City." Similarly, although it is true that Chicago is a collection of buildings, roads, stores, and so forth, it is also true that Chicago is the mechanism by which much of the American Midwest was commodified. If we seek the usual demographic information, we are asking about the former; if, like the historian William Cronon in *Nature's Metropolis*, we seek to understand the relationship between a great city and its hinterland, we are asking about the latter. What is happening, simply, is that our query to the system calls forth from an underlying world—complex yet real—a particular structure that is responsive to that particular query. It isn't that the structure that we call forth isn't "real"; it's that the structure is also, and necessarily, partial. And the means by which that partiality is delineated, our query, is necessarily subjective; the query is based on our purposes and our intent. The partial reality that we call forth is given to us not by the external system ("the real world"), and not solely by our framing of a particular query or context (which cannot ignore "the real world"), but by the interaction between the two. The system itself always remains more complex than what one is are able to capture at any particular time, with any particular perspective.

That the technology systems we have been speaking of are completely built by humans does not mean that their dynamics and their evolutionary paths are planned, or determined, by humans. That we compose them with Level I technologies does not mean that they don't operate at Level III. The Internet, for example, is completely built by humans using Level I technologies: optical fibers carry information, routers send packets hither and yon, and computers access it, all as intended. But insofar as the Internet supports the evolution of synthetic realities and the metaverse, creates information structures that dramatically change policies and politics around the world, builds heretofore unimaginable social networks, and becomes a major domain of *sub rosa* conflict among world powers, it is neither understandable nor transparent. When humans design themselves using various enhancement technologies, they will not make themselves just as they please; the human system—scaled from the individual mind through community, polity, society, religion, and ideology, and interacting with the surrounding physical and information worlds—is simply too complex for us to bend to our parochial ambitions and perceptions.

Think about how you think. You begin with an assumed context that you are seldom even aware of, then work out the details of the problem that you want to ponder within that context. But the continuing evolution of information and cognitive technology systems, in combination with the accelerating fragmentation of time, space, and culture, dramatically decreases the stability and the universality of all cultural constructs. The assumed context can't be considered stable. As the railroad example suggests, this phenomenon is not new; however, the rates of complexification and instability may have increased so much that it appears in a qualitatively new guise, and there isn't enough time for even the transient stability that allows institutions to catch up and adapt. Law and regulation, for example, fall further and further behind the technological frontier, offering neither protection nor guidance as new technologies,

services, and social practices evolve. Society fragments as unprecedented growth in information enables individuals to construct their own self-reinforcing world: you live in a *New York
Times* space, I live in a Fox News space, and fundamentalists
of all stripes build their own communities across the Internet.

This increased fragmentation has two profound effects: not
only does it render the social and cultural landscapes onto
which we gaze more unstable; it renders that which gazes—the
self and our individuality—more contingent as well. The conspicuous increase of fundamentalism across many belief systems and in many societies reflects, in part, an effort to create
a stable ground. But such ground is increasingly unstable, as
the diversity of fundamentalist refuges itself so clearly shows.
Meanwhile, those in the global elite embrace integration with
their surrounding technologies, which helps them master the
continually shifting techno-human landscape. Of course they
are transhuman—they always have been. While one way to
manage this evanescence is through reversion to absolute so-
lipsism and moral relativism, we see in this tendency an invita-
tion to nihilism. For us, the challenge is different: to ensure that
our mental models and cultural constructs are adaptive—to
embrace, and manage, our contingency without cutting ourselves entirely loose from our cultural, political, and intellectual moorings.

An important effect of this contingency becomes evident
when we evaluate our usual assumptions about the relationship
between humans and technology. Here, again, "transhumanism" is a valuable talisman, for it is often used as a code word
for a particular way of framing the interrelationship between
humans and technology. For proponents, "transhumanism"
seems to be almost a *deus et machina* that waves humanity into
a paradisiacal future; for opponents, it seems to be emblematic
of some sort of victory of technology over the human, as if each
were a separate domain. Both of these worldviews, grounded
as they are by the mythical categories of the Enlightenment

(individuals, mind, nature, and so on), and backed by resort to the Enlightenment's chief weapon of applied rationality, fail to grapple with the radical contingency of the techno-human condition. The human and the technological will not clash, with one or the other emerging victorious. Nor will technology, reaching down its empathic paw, raise us from the trials and tribulations of being human. Rather, what will happen is what has already been happening: the two will continue to merge and re-make one another on the individual scale, on the institutional scale, on the social scale, on the planetary scale. Printing presses and books created a new kind of religious scholar; the Internet and Google create a new kind of student. Such changes may be profound. New human varietals will continue to emerge—indeed, "digital natives," comfortably embedded in their ICT networks, may already exemplify this evolution. And integrating powerful new technologies into society may—and probably will—entail substantial damage, much as the development of the printing press enabled widespread study of Christianity in Medieval Europe in settings not controlled by the Church, which enabled the Reformation, which in turn played a part in hundreds of years of bloody religious warfare. So, the question of the relationship between the human and the technological is not resolvable by the existing and coherent (and therefore partial and at least partly wrong) worldviews now battling over "transhumanism." Instead, we have a state of radical and irreducible contingency that requires—if it is to be comprehended sufficiently to allow effective engagement with its implications—simultaneous contemplation of many different and perhaps conflicting worldviews.

Thus does the anthropogenic world resulting from the Enlightenment, and its industrial and scientific revolutions, seem to create a strong and necessary commitment to the development of internally consistent and coherent worldviews and ideologies, even as it demands the philosophic flexibility necessary to respond to complex systems unrolling in unpredictable and uncertain majesty. In short, in a complex world, the intelligible and the rational may often conflict; Level III rationality—a capacity to link cognition to desired outcomes in the world via action—can only emerge from a commitment to confronting and working with (we would say "managing," but it isn't clear that we can actually do that in any strong sense with such complex and powerful systems) that which is incomprehensible.

Might this irreducible complexity and contingency itself favor certain societies over others? Simple cause-and-effect suggestions obviously do not pertain here, but one can ask some interesting questions. Is a society that embraces some aspects of transhumanist technology more likely to be able to integrate humans and technology as rates of technological evolution accelerate? In Japanese anime, for example, human/technology combinations are often the good guys, and in Japanese society humanoid robots seem more comfortably accepted than in the West, where Frankenstein myths remain powerful. Might cultures that encourage multiple visions of an underlying complex reality have a long-term advantage in a period of rapidly destabilizing technological change? (Hinduism allows many different gods and goddesses, many of whom have additional representations—Krishna as an avatar of Vishnu, for example—and all of whom represent different aspects of an underlying reality understood to be too complex for direct access.) Obviously such cultural tendencies are not deterministic; they do not substitute for strong governance and effective economic institutions, for example. But we wonder whether cultural characteristics such as these might enhance cultural competitiveness in the long term, at least at the margin.

We aren't finished; it gets harder.

Not only can no one understand "the whole," but a good worldview or ideology is often just what is necessary for simplifying things enough to enable individuals to operate in the real world without losing any sense of meaning. Such digests of reality define and engage communities, turning complex questions of dynamic interaction into psychologically and politically manageable choice scenarios and symbols that simultaneously reflect and validate particular views of modernity. (One thinks, for example, of President George W. Bush's division of the world into pro-terrorism and anti-terrorism camps, or the tendency of environmental ideologies to characterize the actions of individuals as either "good" or "bad" for the environment.) Creating such frameworks to provide structure in a complicated world has, paradoxically, been one of the strengths of the Enlightenment. As Robert Conquest points out, the ideologies that are most familiar to moderns are in fact products of the Enlightenment, often displaying a characteristic integration of applied reason and millenarian utopianism.[7] Indeed, the recognition, by Herbert Simon and other scientists who study decision making, that rational action requires simplification and filtering—"bounded rationality," in Simon's term—counts as one of the key psychological insights of the twentieth century.[8]

But, like other elements of the traditional Enlightenment, simplifying worldviews or ideologies are continually rendered obsolete by their very success. If simplification is a necessary mechanism for cognition and for day-to-day functioning, it is nonetheless problematic if the essence of the environment is, in fact, its complexity. This difficulty is magnified because worldviews and ideologies are necessarily based on assumptions and analyses that derive from past experience, and thus have been rendered obsolete by rapid and discontinuous change. Whether you are creating a scientific model or a political ideology, you determine what is important to include and what can be ignored; you then build a structure based on what is important for capturing the essence of the situation. But you can do this only because you have some pre-existing idea about what matters, so that you have a principled way of knowing (where the principles may be scientific or moral) what to keep, and what to ignore. If your ideas about what matters are obsolete ("Economies must always grow!" "Housing prices can never go down!" "Science can solve the problem!" "Think globally, act locally!"), then the structure you create will also be obsolete, and it will be ineffectual in helping you to navigate the environment in which you find yourself.

Simplifications such as ideologies also cut off information transfer and dialog, and thus, in a complex environment, may cut off exploration of new ways of understanding the world. Jared Diamond points out that the Christian and European identity clung to by Scandinavian settlers in Greenland initially helped them thrive but ultimately prevented them from considering and adopting Inuit customs—a closure of cultural and practical options spaces that ensured their doom as climate conditions changed.[9] And, in a way, the belief systems that the proponents and opponents of transhumanism bring to their dialog only prove the point: they hide, rather than explore, the real implications of emerging technologies and their effects on humans and their communities.

It is not, then, that coherent belief systems such as ideologies are bad (although many of them seem to be bad in application, as any familiarity with the twentieth century would confirm); rather, it is that, in a period of rapid, discontinuous, fundamental, global, multicultural change, coherent belief systems are an obstacle to the effective structuring of comprehension and action. Because ideologies with a quasi-rational and thus Enlightenment mien have become a convenient way of simplifying a complex environment, their failure not just in practice but also in principle is a further weakening of the original Enlightenment project.

Again, we do not argue for relativism. Some worldviews (e.g., German National Socialism and the later Maoist and Pol Pot versions of Marxism) are obviously and always unacceptable; others (e.g., the Enlightenment commitment to rational action based on evidence and induction) work well at Level I, and fairly well at Level II, where the connections among goals, technologies, and social and cultural context are often visible.

At Level III, however, all worldviews, even the most privileged, such as the science discourse and liberal democracy, are partial, and a failure to explore different worldviews and identify appropriate options can rapidly become ineffective or even fatal. So long as the Greenland weather behaved like European weather, the Christian and European cultural worldview served the settlers well: they were in a Level I and Level II world. But when climate changed, they were thrown into a Level III situation—highly unpredictable and contingent—and failed to adjust.

The intellectual confusion that occurs when one applies Level I and Level II coherent worldviews to a Level III condition is quite evident today in the climate-change arena and in the infatuation with "carbon footprints."[10] For example, a professor writing in the *Medical Journal of Australia* recently called on the Australian government to impose a carbon charge of $5,000 on every birth, to charge annual carbon fees of $800 per child, and to provide a carbon credit for sterilization.[11] Articles in *New Scientist* have suggested that obesity is mostly a problem because of the additional carbon load it imposes on the environment,[12] that a major social cost of divorce is the additional carbon burden resulting from splitting up families, and that pets should be ~~should be~~ eliminated because of their carbon footprints ("Man's best friend, it turns out, is the planet's enemy").[13] A recent study from the Swedish Ministry of Sustainable Development argues that males have a disproportionately larger impact on global warming.[14] ("Women cause considerably fewer carbon dioxide emissions than men, and thus considerably less climate change.") The chairman of the Intergovernmental Panel on Climate Change said that those who downplay the potentially catastrophic consequences of climate change are no better than Hitler.[15] (He now claims that his words were taken out of context, but the reporter who conducted the interview, Lars From, stands by it.) E. O. Wilson has called such people parasites. The columnist Ellen Goodman has written that "global warming deniers are now on a par with Holocaust deniers."[16]

There are always fringe articles and unfortunate comments in areas of active public debate. But the sheer volume of articles, the vicious language, and the retranslation of so many complex social and cultural trends—divorce, obesity, gender conflict, child-bearing—into terms of carbon footprint suggests that something more fundamental is going on. Most obviously, the extreme language—comparing academics who disagree about interpretation of data to Hitler, or to Holocaust deniers—is indicative of a profound if subtle reframing of climate change. One does not debate Hitler. The use of such language indicates a shift from helping the public and policy makers understand a complex issue to demonizing disagreement, especially disagreement about policies favored by some representatives of the scientific community. The data-driven and exploratory processes of science are choked off by inculcation of belief systems that rely on archetypal and emotive strength. Importantly, the extreme language is directed not against those who deny anthropogenic climate change completely, of whom there are few left (a credit to the traditional scientific debate process while it still existed in this area), but against those who, while accepting the existence of the phenomenon, remain uncertain about the timing and the severity of the problem, or about how best to confront it successfully. The authority of science is not a foundation for factual enlightenment but an ideological foundation for authoritarian policy prescriptions that might otherwise be difficult to implement and which, in a Level III world, are more likely to deliver surprises than solutions.

More and more articles and proclamations, some verging on self-parody, seek to redefine social and cultural phenomena in terms of carbon footprint. On one level this expansion of explanatory skeins is trivially valid: life is based on carbon, so, to the extent one participates in life, one will inevitably affect the carbon cycle. But defining complex human behaviors and states—such as obesity, or having children and pets—in terms of carbon footprint begins to create a new structure of good and evil in society. Obesity is now questionable not for reasons of health or Calvinist theology, but because obese people are destroying the world through the carbon-footprint-expanding sin of gluttony. A complex public health problem is nicely converted into a simplistic moral mapping by jamming a Level III system into a Level I simplicity: "Carbon footprint—wrong!" Similarly, the report of Sweden's Environment Advisory Council uses climate change to reinvent the eco-feminist condemnation of males as evil destroyers of the environment. (*New Scientist*'s news item on the report referred to "male eco-villains.") The campaign to create a moral universe predicated on carbon footprint, which began with initiatives against sport-utility vehicles, is now extending across society as a whole. Climate-change science and climate-change policy are rapidly becoming carbon fundamentalism, a simplistic but comprehensive structure of moral valuation that can be applied to virtually any individual or institution.

This moral evolution bears the fingerprints of the Enlightenment because science has been invoked—both by scientists and laypersons—as the key source of information, guidance, and truth underlying the authority of carbon fundamentalism. As we write these words, however, the predictable backlash is gathering momentum. E-mail exchanges among climate scientists, and errors in the reports of the Intergovernmental Panel on Climate Change, have provided ammunition for those who oppose carbon-footprint fundamentalism to use in attacking the authority of the science.[17] Public support for action on climate change is declining not just in the United States but in many European countries as well.

In some ways, then, the climate-change discourse is not a brave assault of knowledge against ignorance and greed, but rather is yet another example of how Enlightenment rationalism—the foundation for linking rationality to action for 300 years—is failing. Just like transhumanism, the climate change discourse relies on heroes and villains that spring from our past, but fail in our future. Many other issues—immigration, the "war on drugs," international terrorism—have been subjected to a similar application of Level I worldviews to Level III problems. The modern international criminal sector, a trillion-dollar industrial machine, is funded primarily by rents that can flow only so long as drugs are illegal. The costs of keeping drugs illegal—narco states, human misery, violence—are huge and in countries such as Mexico may pose a threat to the state itself. Yet these costs are justified by worldviews—"Just say no!"— that, like the one that characterizes the obsession with carbon footprints, are staggeringly naive and superficial. And yet, as was the case in our earlier critique of the benefits of enhanced intelligence, we note that some of the best and the brightest among us—neocons, climate scientists, policy wonks—are unable to resist the political seduction of Level I language, and its promise of cultural and political authority.

should be
a higher change takes too long.
level ?? ? ?? ? ? ? ? ? ? ? ?

7
Killer Apps

In order to further clarify our approach, as well as to explore a system deeply implicated in transhumanist technologies, we now want to bring our analysis to bear on a specific area of technological application. In particular, we want to make clear that human enhancement and technological complexity are not just boutique playgrounds for the techno-elite and its observers, but in fact lie at the very heart of the most powerful driver of innovation and social transformation: the rapidly evolving interplay among emerging technologies, military operations, and national security. The intimate relation between technological evolution and military activity appears to be central to the techno-human condition. From the Trojan Horse to the longbows of Agincourt, from ships of the line at Trafalgar to railroad lines in the Civil War, from World War I machine guns to World War II tanks, from the nuclear annihilation of Hiroshima and Nagasaki to the "shock and awe" of Iraq, one version of human history is a technological telling in which weaponry and military victory march in lock-step.

The combination of the existential challenge to society represented by warfare and the immediate advantage over an enemy that new technologies can deliver has been one of the most important accelerators of technological innovation and diffusion through history. This tendency has often been countered by the deep conservatism of military institutions and personnel—a conservatism justified by the high cost of any unsuccessful military innovation. (If you get it wrong, you die, and maybe your country dies too.) Yet the costs of failing to innovate when facing a challenging enemy are even greater. Moreover, innovators may be inclined toward aggressiveness: if you think you are technologically superior to your opposition, you may well initiate conflict more readily.

The relationships between military-technology and security-technology systems, and the consequent social and ethical issues and changes, are at least as complex as anything we have discussed so far, so understanding and managing them in ways that are likely to enhance long-term military advantage and security is by no means a straightforward, Level I task.

Consider something as simple as gunpowder. An innovation based on relatively simple chemistry, gunpowder was first developed in the eighth century during the Tang Dynasty in China. The substance—first used as a treatment for skin diseases and a fumigant—eventually became a military technology, but for a long time it was used only as a display weapon, full of sound and fury but signifying rather than imposing power. That's because, as a mixture, it generates a relatively low-velocity explosion, and it has a tendency to separate out into its constituent chemicals: sulfur, charcoal, and potassium nitrate. This explains the importance of "corning," a technique, developed in Europe in the early 1400s, in which the dry mixture was wetted, then dried and milled, yielding stable particles of gunpowder. This advance stimulated the development of better weapons (which in turn stimulated the science and technology of metallurgy), which in turn enabled soldiers to defeat armored knights, with the effect of shifting power between classes. It took a lot of resources to support an armed knight, so knights were, of necessity, members of the elite; any peasant, however, using even the first primitive corned-powder hand weapons, could be trained to be an effective military presence. Power also shifted among cultures. Gunpowder equalized the military capabilities of European armies and the Steppe peoples, whose fierce cavalry armed with bows no longer posed as great a threat as they once did. But such relatively obvious effects hide many other, more complicated patterns. Gunpowder weapons created economies of scale in warfare, because cannon and powder, and platforms such as galleons, and the foundries and shipyards necessary to produce them, were not cheap. Nobles couldn't compete with Renaissance princes and centralized states, and finance became the key to national military greatness. Moreover, gunpowder armies required complex logistics: a pre-gunpowder army could live off the land, and in fact that was accepted practice (although difficult for the serfs involved, of course). But farmland does not produce heavy shot, or gunpowder, so managing the supply chains of moving gunpowder armies became much more complicated, requiring new competency in military bureaucracies that managed logistics. Gunpowder armies required drilling to function effectively; thus, organized military training, and a regimental system that institutionalized a more formal military structure, became important military advantages.[1]

All this innovation and cultural adaptation led to the evolution of a more professional military; the "come as you can" feudal military organization (think of the Crusades) was rapidly superseded. Moreover, a prince could field only a limited number of knights, but gunpowder hand weapons could be made fairly inexpensively, so peasant armies became possible. And gunpowder cannon mounted on European ships made short work of coastal defenses in India, in the Spice Islands, and in the Caribbean, contributing to increasingly centralized power, and to empire. (In fact, historians use the term "Gunpowder Empires.") In short, a military technology that couldn't be much simpler (three chemicals and a spark), and which for a surprisingly long time played only a marginal role in military encounters, ended up restructuring the world in fairly short order. Like the railroad, and like the early textile factories, this isn't something you would have predicted *a priori*. But a simple technological step (corning technology probably being the crucial advance) resulted in a seismic shift in regional and global cultural and economic patterns. Those who took advantage of the modest innovation did not do so because they had superior prior knowledge; they did so because they studied, learned, and experimented. They muddled, and it is an important observation that some muddling is more effective, and more informative, than other muddling. The success of the Swedish military in the 1630s against far larger states, and Prussia's success in the 1860s, were not just stochastic, but careful, inspired, and opportunistic muddling. Max Boot, in his book *War Made New,* makes this point nicely in summing up his discussion of 500 years of war: "There is no rule of thumb to suggest how much or how little a military should change in response to technological developments. . . . History indicates that the wisest course is to feel one's way along with careful study, radical experimentation, and freewheeling war games."[2]

Such historical musings provide some background for appreciating the scale of social and cultural change that could soon occur as a result of rapid technological change driven by military necessity. Emerging technologies are likely to have destabilizing effects within military itself, potentially affecting not only military operations but also military culture and organization and the broader social perspectives on military initiatives and missions. The latter point may not seem worthy of much concern; however, because the military culture and military institutions are the carriers of the ethics and rules that govern war, from following international agreements such as the Geneva Convention to the powerful idea that officers are personally responsible for what happens under their leadership, cultural change can have a dramatic effect on the conduct of warfare, and especially on how combat affects civilian populations.

Thinking about National Security as a Level III System

The relation between military application (a Level I and Level II problem) and national security (Levels II and III) provides an entry point for thinking about technological evolution and broader questions of concomitant institutional and social complexity. This exercise, in turn, provides a basis for a more grounded understanding of how technology systems play across different levels of complexity.

To begin with, we tend to view military operations as chronologically bounded, whereas we tend to view national security as an ongoing concern. For example, most people would agree with the claim that Japan and Germany "lost" World War II—as indeed they did, militarily. But those countries are now wealthy, with democratic governments and citizens who are among the world's most productive and best educated, and both are strong pillars of the West. Similarly, few doubt that the French, and then the Americans, "lost" Vietnam, yet that country is now an increasingly important trading partner of the United States, and, though not democratic, it certainly poses no challenge to American power. What "hard power" (military combat) appeared to lose, "soft power" (trade and market-economy forces, cultural attractiveness) achieved. History does not judge "win" or "lose" nearly as quickly as combat does, and a systemic view of security must be careful not to jump to conclusions.

Indeed, the idea that national power is predominantly a matter of military capability has long been recognized as simplistic and naive. Modern hegemonic power emerges from at least five sources of dominance: economic, scientific and technological, military, institutional, and cultural. Economic power linked to manufacturing prowess, for example, propelled Japan into the first rank of powers after World War II despite its relative military weakness (which was compensated for by the American military "umbrella"), whereas institutional inadequacies in the political and financial realms have contributed significantly to its drift in recent decades. The European Union retains its dominant international position, despite an unimpressive military, in part because its institutions reflect a resilient and open social-democratic political framework and its culture is widely admired. China's rise reflects an economic boom and extraordinary innovative energy, and those who predict future decline do so largely because of perceived institutional weaknesses. Even in the case of past empires, flexible governance structures and institutions that knitted together distant lands were as important as sheer military power. Roman roads and administration, for example, did as much to support the Empire as the Legions, China's famous Confucian bureaucracy has throughout its history been a unique unifying strength of that culture, and England's domination of the globe from 1815 to the early 1900s owed as much to superior sophisticated financial and organizational innovations as to the English Navy—as previously had been the case with Holland. (However, none of these powers would have lasted long without underlying military muscle, which created the environment within which civil institutions could operate safely.)

In a world where the most advanced economies are characterized by increasing reliance on information networks, by highly flexible economic and political institutions tending toward the flat and virtual rather than the hierarchical, and by reduced dependence on direct control of resources, the key to obtaining and keeping hegemonic power increasingly is balance among the five sources of dominance listed above. Certainly this seems to be the case with the United States, which until recently has been the one power that has appeared to be globally competent in all five domains: the largest economy, unmatched capability in science and technology (supported by equally preeminent academic and high-tech industrial infrastructures), a military that is far more advanced than any other in many ways, an institutional structure that is relatively transparent and is defined by law rather than relationship, and a cultural ascendancy that is reflected in the widespread idea (in the United States, at least) of American exceptionalism, in entrepreneurism, and in brand dominance (especially in consumer culture: Coca-Cola, McDonald's, Disney, and so on). Although different actors react to different aspects of this hegemonic power structure—the French, for example, who do not fear the United States militarily because it is an ally, tend to express concern about American cultural domination, while the Chinese, who face America's projected power (particularly its blue water Navy) as they look toward Taiwan, are much more concerned with the military domain—it is the balance of high competence in all five sources of dominance that has made the U.S. formidable.

But the essence of American power is subtler than just the components in themselves, as the experiences of Japan, China, and the European Union show. It is not enough to develop a more aggressive and successful economy (as Japan in some ways did and as China is doing) or to challenge American exceptionalism and cultural appeal (as Europe does). To truly challenge America over time, a rising power must become competent in all five sources of dominance. To complicate matters, these domains are not independent of one another; rather, success in each requires synergism among all of them. A challenger, then, need not have the same balance as the United States, nor need it make the same choices, but it must be able to compete successfully, at the global scale, in all five—and to integrate them effectively so they will be mutually supporting. Thus, while U.S. strength across all five areas will be difficult to maintain, other nations will not easily achieve the necessary multi-dimensional competence, even now when economic power is arguably shifting eastwards from America to Asia, and south to Brazil and perhaps someday Mexico.

Consider the daunting task faced by a country trying to match the United States in science and technology (S&T). American investment in research and development by the government and private industry accounts for about a third of the total global investment and ensures American supremacy in this critical area. But, worse yet from the perspective of those who would challenge America's S&T capabilities, these strengths are further buttressed by institutional and cultural dimensions. The American system of higher education is considered by many to be the best in the world overall; moreover, it pulls in intelligence from other societies around the world, much of which remains in the United States either as intellectual property or as working, thinking, highly educated human beings. The American venture-capital system, again the most developed in the world, supports this structure by ensuring that advances in science and technology are rapidly translated into entrepreneurial activity and thus economic power. The American military has been one of the world's most important customers for emerging technological capabilities, which then smoothes the way for commercial introduction. The American culture, which tends to be technologically optimistic, underpins these systems. A culture that seeks to match American supremacy in science and technology, therefore, cannot do so simply by increasing research spending, or by trying to develop a few world-class technical institutes. It must create a network across its culture that understands excellence in science and technology as an emergent characteristic of excellence across all five sources of dominance—a much more difficult task.

Concomitantly, any erosion of American security is unlikely to be the result of a purely external threat. America after the Cold War had a dominance that, although displayed in terms of the hard-power supremacy of military and economic might, was also hegemonic in its cultural dimension: the United States attracted brains and capital, and, indeed, seemed to embody the mythic qualities of its exceptionalism because of its uniquely open, optimistic, entrepreneurial and mobile society. Its real strength was its "brand," which reflected and required economic and military success but was not limited to those domains. This strength was evident in the number of students attracted to American universities, in the number of non-native-born entrepreneurs who created Silicon Valley and its Texas, Massachusetts, and Oregon counterparts, in the success of American consumer goods and cultural exports (films, games, the Marlboro Man), in the continued attraction of the American experience for people persecuted or discriminated against in other cultures, and in the success of global market mechanisms such as the World Trade Organization. The real key to long-term American dominance, in other words, was a larger-than-life projection of American values and culture across the world.

In this light, the decision to invade Iraq in 2003 undermined long-term American power, in part because it conflated technological dominance and military power at Level I (ships, planes, smart munitions) with national security at Level III. (The objective of fundamentally restructuring Iraq, and Middle Eastern governance systems in general, which was also claimed as a rationale for military action, similarly conflates Levels I and III.) America decided to build a technological Maginot Line, and a very expensive one, in a world of cultural blitzkrieg. Its leaders believed too much in Level I technology—"shock and awe"— in a situation (Middle East politics, religion and history, combined with oil) that was clearly Level III. And so in May 2003, President George W. Bush trumpeted the conventional military victory in Iraq during a photo op on the deck of an aircraft carrier, with a banner famously declaring "Mission Accomplished" adorning the vessel's bridge. The carrier was perhaps the ultimate symbol of Level I technological functional effectiveness. But the president, it turned out, was really declaring a triumph of category confusion.[3] The real war was only just beginning.

Category confusion also characterized the response of the American political system to the very real threat of terrorism. Level I technologies and approaches—such as enhanced scanning at airports and, on the social side, increased surveillance and stricter regulatory controls on individuals—were avidly introduced; consideration of Level III issues, such as the linkages among culture, status, current political events, and long-term conflicts between worldviews, was not only discouraged but even labeled traitorous. Initial responses to relatively undefined challenges may, of course, be granted charitable dispensation against too much second guessing; that such initial responses still form the basis of the social response to terrorism a decade later is less encouraging.[4]

The Iraq experience extended the gestalt of Level I technology (direct and explicit integration of tools and goals) to a domain (the exercise of soft power) where it became counterproductive. Certainly the massive deployment of technology was extraordinarily successful in a Level I sense—America invaded Iraq and won all the battles. But invading Iraq, or indeed any such invasion, is a Level III activity. The Iraq mission failed because a war to change a traditional culture into a modern democracy is not a Level I exercise in military technology; it is a high-risk, high-uncertainty exercise in trying to manage Earth systems—a serious Level III challenge. Indeed, even the idea of employing a Level I technology on a battlefield to kill an enemy no longer seems to be as simple as it once did. Where is the battlefield in Afghanistan? Where was it in Vietnam? Who, even, is the enemy in such conflicts?

How deep does this rabbit hole go? Pretty deep. Figure 7.1 provides an initial idea of the Level I, Level II, and Level III terrain of emerging technologies, military operations, and national security, and illustrates the dramatic complexity of the techno-security challenge. Several things are immediately obvious. First, concepts and assumptions that have been stable for centuries are increasingly contingent. Second, the system is complex and unpredictable. Third, the Level I capabilities of military technology are not going to get us very far in trying to understand their implications for national security. We'll need a different approach.

The figure identifies four major realms of coupled change. (We've drawn these as if they are independent, but that is for exposition only; in reality, it is because they are not independent that they are so difficult to figure out.) Start with the realm of Revolutions in Military Technologies (RMT); this is the realm of Level I technology, several specific examples of which we will discuss below. Yet, as we have been seeking to make clear, even their functionality as Level I technologies cannot be understood in isolation, but only in the context of an entire frontier of related emerging technologies—just as the reliability of the jet aircraft cannot be understood outside the Level II systems context of the air transport system. And it is this context that gives military and security types headaches. The challenge, that is, goes far beyond developing and deploying telepathic helmets—a plausible enough task, in all likelihood, for the brainiacs at the forefront of cognitive engineering, eventually. And integrating those technologies into a Level II system of battlefield tactics is likely to prove tractable. But what happens when the rest of the world doesn't play by the rules and assumptions that the brainiacs and tacticians are designing for (as it never really does)?

The second realm is Revolutions in Nature of Conflict (RNC). For example, one of the long-standing principles of international relations since the rise of the nation-state has been the absolute right of the state to do what it wants inside its borders. (Indeed, this was the essence of the Treaties of Westphalia, which in 1648 established the international order characterized by nation-states: the King got to decide the religion of his realm, and other states would recognize that decision—*Cuius regio, eius religio.* In a Europe torn by religious conflict, this was a big step forward.) But in recent times, as states have systematically annihilated minorities—in Bosnia, Sudan, Rwanda, Congo, and elsewhere—strong support has emerged for the principle of "responsibility to protect." Under this principle, states can justifiably intervene in the internal affairs of

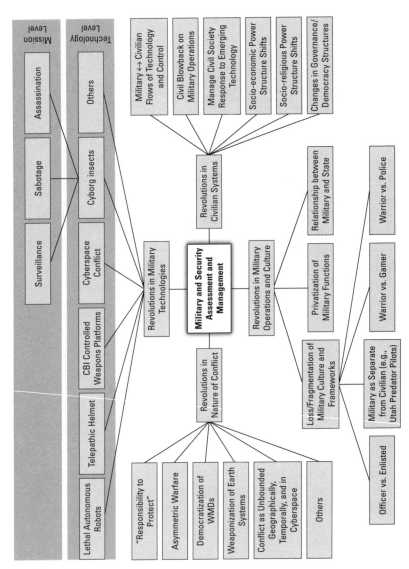

Figure 7.1
The complex landscape of modern conflict. Despite the neat boxes, the complexity and unpredictability of this system derive in large part from the fact that all these issues, technologies, trends, and systems are interconnected and constantly co-evolving.

other states. Combat itself has also shape-shifted—from the choreographed battles of the eighteenth and nineteenth centuries in Europe (from which much of the Western theory of warfare is derived), to the insurgencies of Vietnam and Malaysia, to counterinsurgency as the modern model of conflict. And military combat has morphed into enhanced policing and state-building. Unlike traditional combat, conflicts are no longer limited in time or space. They also involve new players, such as networked non-governmental organizations, that cannot be controlled or attacked through traditional means. In part, this expansion and diversification reflects the success of developed countries, particularly the United States, in achieving overwhelming dominance in traditional methods of war. States and non-state actors like al-Qaeda that cannot compete under the old rules will, quite rationally, seek alternatives: asymmetric warfare. Level I and II become entwined: weapons developed for combat, such as robots and biological enhancements, can be rapidly adopted by such non-traditional groups and used in conflict against their original developers. For example, in Afghanistan, software meant to monitor targets for U.S. unmanned aerial vehicles (UAVs) has been employed by insurgents to monitor U.S. troop positions.

Much of the action in the RNC realm is taking place in cyberspace, often invisible to the public unless some group (perhaps a military organization seeking additional funding for cyberspace activities) has a motive to make a splash. The major players—Russia, the United States, Europe, China, India, and others—probably have compromised one another's Internet systems to the point that a major conflict involving any of them would immediately result in tremendous damage to the states involved. (The mechanisms for introducing such chaos would not be the relatively clumsy denial-of-service attacks that we saw in Russia's assaults on Estonia and Georgia, but rather the "back doors" planted in Internet systems, which would be activated in such a way as to cause appropriate damage in case of attack—"appropriate" being the level of damage required to achieve one's strategic goals under the conditions as they unfold, which need not involve full implementation of "cyberwar" capability.⟩Details, of course, are classified, but it is reasonable to expect that, even now, a new "balance of terror" based on "mutual assured destruction" is evolving on the cyber battlefield. And in the realm of asymmetric warfare, cyberspace is also the obvious place for less technologically advanced nations and for non-governmental organizations to go if they wish to inflict relatively undirected damage on a sophisticated opponent.

The third realm, Revolutions in Civilian Systems (RCS), is one of the main themes of this book, so we will touch on it only briefly here. In short, you don't mess around with powerful Level I technologies without getting institutional, social, and cultural shifts that you weren't expecting. For example, consider the potential for civil blowback arising from technophobia. Suppose some country's army introduces cyborg insects—some sort of robot or cyborg of insect size and function—as a powerful counterinsurgency surveillance tool, and they leak back into civil society and are misused, perhaps by a rogue military contractor. Suppose that society reacts with a ban on such devices, perhaps combined with constraints on research, and that as a result a useful and potentially desirable technology is eliminated from that country's military repertoire (though not, of course, from its opponent's).[5] The general idea here is that attention to the dynamics of Level III RCS systems could help protect a broad social good, such as privacy, and thus also protect the state's Level I ability to develop and use important military technologies and to accomplish military missions.

The fourth and last realm, Revolutions in Military Operations and Culture (RMOC), encompasses the potential for emerging military technologies to change the underlying foundations of the military itself. This potential is not just a military concern. Consider that military culture is one of the reasons why some countries are prone to coups and others are not. We are entirely on Level III turf here. For example, the military culture in most developed countries assumes a professional, highly trained core around which volunteers or draftees are assembled. But operations in Bosnia and Afghanistan are more like policing operations, aimed at protecting civilians and the built environment rather than blowing it up and looking for bad guys. Policing and combat require very different kinds of training, and very different institutional cultures. Good soldiers are seldom good policemen, and certainly not both at the same time. Technology ramps up the complexity further. In his book *Wired for War*, Peter Singer reports that the United States had no ground robots when it invaded Afghanistan in 2002, 150 of them by the end of 2004, 2,400 by the end of 2005, 5,000 by the end of 2006, and 12,000 by the end of 2008. The number of UAVs has increased even faster. One result of this escalation is a profound culture conflict as officers who absorbed traditional military culture as they worked their way through the ranks find themselves competing with gamers, and as military personnel who are in active combat situations in the Hindu Kush are supported by remote-control UAV pilots who go home to their houses in American suburbs when their day is done.

On top of that, many functions once carried out by active-duty military personnel, including such combat roles as guarding convoys and running military prisons in locations outside the United States, have now been turned over to Halliburton, Xe, and other private companies. (Blackwater Security was renamed Xe after some unfortunate incidents involving the shooting of civilians.) From the perspective of emerging technologies, outsourcing the state's monopoly on military action has huge implications: it would be one thing to have cyborg-insect technology under state control, but quite another should it leak into civil society through quasi-governmental private firms, with neither the culture nor some of the legal protections that might otherwise apply.

Our schematic mapping of RMT, RNC, RCS, and RMOC is meant to illustrate two larger points. First, each of these realms is changing, each is contingent, each is unstable, and each is coupled to the others; taken together, they form a potent Level III system. Second, rapidly emerging military technologies, particularly technologies that expand and extend human cognitive capacity, are being injected into this volatile context. The combination of contextual instability and rapid technical change creates at least the potential for radical transformation and even destabilization of economic, social, and technological systems—not to mention military institutions and capabilities. This potential is especially potent because technologies that provide additional military capability or help protect troops tend to be regarded as important and worthwhile for that Level I reason, and thus there is a strong incentive, especially if benefits are clear, to deal with more hypothetical or less pressing Level II and Level III effects later.

To explore these dynamics further, let us look at three types of emerging military technologies. These examples are meant to illustrate with some specificity how one might think about a technology in terms of its Level I, Level II, and Level III implications. Our choice of military technology is also meant to focus attention on the domain of human endeavor in which significant technological enhancement of humans is likely to play out soonest, at the highest rate, and with the most far-reaching consequences.

Caution: Remember that these technologies are still under development. Though some of the preliminary research looks promising, they are best regarded as scenarios rather than done deals. On the other hand, our descriptions are based on public sources. If there have been significant advances on the spook side, we wouldn't know that (or if we did, we wouldn't tell you).

Case 1: Cyborg Insects and Other Surveillance Devices

At present, this suite of technologies includes two main categories. The first involves implanting electronics in real insects; the second involves the creation of robots with the size and functionality of insects.

As of 2009, researchers at the University of California at Berkeley had successfully implanted electrodes and a radio receiver on a giant flower beetle, allowing control of the beetle in flight. Researchers have also implanted moth pupae with silicon chips that are retained by the moths as they pupate. The U.S. Army has funded work on the development of mechanical bugs for use in battlefield environments, particularly in ISR (intelligence, surveillance, and reconnaissance) applications. Research is not limited to individual robot or cyborg units, but includes efforts to link different cyborg-insect units into grids, thus dramatically increasing their computational and surveillance capability. In this, cyborg-insect technology—as with many technologies that appear new and radical—is best understood as an extension of existing research trends. In particular, BAE Systems, a military contractor, had previously developed a technology, called Wolfpack, consisting of networked unattended ground sensors that, in BAE's words, provided "a complete end-to-end ground sensor system consisting of remote sensors, advanced detection, tracking and jamming algorithms, and controller workstation, capable of integration into a larger C4I system."[6] Not only do such sensor systems provide information on battlefield conditions; when coupled to intelligent weapons systems, they provide rapid and precise identification of, and appropriate response to, targets.

While scientists have focused much attention on flying platforms, the cyborg-insect zoo under consideration by military planners and research teams includes variants of spiders, high-jumping insects such as grasshoppers, flying moths and bees with detachable payloads, and insect-size composite devices with a variety of characteristics. Among the considerable technical challenges that remain are providing energy for such small platforms (especially as guidance, surveillance, or attack modules are incorporated in the payload) and miniaturizing functional equipment to fit on an insect-size carrier.

Table 7.1 sketches out some of the implications of cyborg-insect technology.

From a Level I perspective, the alignment of technology, goal, and policy is pretty clear: surveillance cyborg insects will be a powerful military tool in a counterinsurgency environment. Moreover, if equipped with offensive capabilities—a fatal stinger, for example—they may have ethical advantages over alternatives such as Predator drones, which often cause collateral damage even when precision munitions are used. A weaponized cyborg-insect system could identify a specific target and then immobilize or otherwise attack just that target, reducing civilian and property collateral damage considerably.

Level II and III are different matters. At Level II, cyborg insects could be a threat to personal privacy should they be deployed in civil society, especially if they are in deployed in military operations carried out by private sector firms. The difficult balance between public safety and security and privacy and independence that most societies struggle to define and maintain could be pushed significantly toward the "intrusive state" end of the scale by such technologies. Given current civilian surveillance strategies such as the ubiquitous video cameras in the United Kingdom, such concerns don't seem paranoid or merely hypothetical.

But if cyborg-insect surveillance technology spreads, Level III implications will kick in. In the ongoing struggle between totalitarian and open government, technologies play unpredictable roles. Twitter, for example, as a means of spreading information about domestic unrest around the world, seems to push toward openness; cyborg insects could push strongly

Table 7.1
Technology-level matrix for cyborg insect.

	Goals and effects	Implications for action
Level I: Military effectiveness	Reduce collateral damage and increase operational efficiency in counterinsurgency operations	Goals and technology align; therefore adopt technology
Level II	Protect civilian populations from terrorists and, through mission creep, from criminals in society at large	Implement technology; but technology alone may not lead to achievement of stated goal
Level III: Social and cultural effects	Ensure orderly society with low risk to citizens and high level of national security. Effect likely to be reduction of privacy and enablement of "soft" or "hard" authoritarian state.	Optimistic goals likely to be undercut as those in power adopt cybersect technology to their own ends; Level I and Level III implications potentially in fundamental conflict

in the opposite direction. Moreover, cyborg insects are a technology platform, not just a surveillance technology—flying a swarm of cyborg insects into a nuclear plant to destroy it would be as easy as collecting data on a rival or an opponent. Cyborg-insect technology also offers capabilities that would be useful to many elements of society, including organized crime, divorce lawyers, parents wanting to track their children, and political parties seeking dirt on their opponents (or, more creatively, using the technology to plant dirt on the opponents). And imagine how ugly politics could get if politicians' media mouthpieces had their own fleets of cyborg insects.

Case 2: Telepathic Helmets

This technology is based on research in which non-invasive monitoring of brain activity or emissions is used to determine what noun, image, or moving image an individual is thinking of, and it is being integrated with greatly expanded computer-based modeling of brain architectures. (IBM announced in 2009 that it had created a simulation where the number of neurons and synapses exceeded those in a cat's brain.[7]) The goal is to create a computer-brain interface (CBI) in the form of a helmet that would read a soldier's brain waves and transmit their content to other members of a small unit, thus enabling the unit to act more as a cohesive whole and to share critical information instantaneously and easily: a telepathic helmet. The U.S. Army is sufficiently attracted by the idea that it is funding a multi-million-dollar exploratory project involving researchers from Carnegie Mellon University, the University of California at Irvine, and the University of Maryland.

Table 7.2 sketches out some of the Level I, Level II, and Level III implications of telepathic-helmet technology. This is a more complicated system than cyborg-insect technology, so it may well take longer to move from R&D to deployment.

Now, it is true that thought-controlled toys are already available, and that headbands that allow gamers to mentally interact with their avatars cost less than $300. But it is a long way (those working on the technology suggest decades) from such devices, which use relatively high-level brain function patterns, to being able to process, transmit, and communicate the more intricate thoughts that an effective telepathic helmet would require. Yet the journey has clearly begun, and it is not too early to begin considering what may lie ahead.

Level I implications, as usual, are relatively easy, both to understand and to evaluate from operational, policy, and ethical perspectives. Telepathic helmets would enhance the performance of small combat units, especially in situations in which face-to-face and voice contact is difficult (army special operations, for example). The technology blends well with, and is synergistic with, the augmented-cognition helmets currently in development.[8] It will thus increase field effectiveness, reduce unnecessary mortality (including from friendly fire), and quite possible reduce collateral damage, especially if combined with other technologies (e.g., if telepathic helmets are linked to cyborg insects, or to augmented-cognition technologies gridded across the battlefield). So at Level I it's a go.

Table 7.2
Technology-level matrix for telepathic helmet.

	Goals and effects	Implications for action
Level I: Military effectiveness	Enhance performance of small units in combat	Goals and technology align; therefore adopt technology
Level II: Adoption in civil society	Protect civilian populations from terrorists and, through mission creep, criminals by, e.g., enabling non-invasive distance capture of thoughts	Implement technology, but technology alone may not lead to achievement of stated goal
Level III: Social and cultural effects	Goal to ensure orderly society with low risk to citizens, and to protect national security. Likely effects: complete elimination of privacy; social instability as subconscious and unfiltered thoughts become explicit; unpredictable restructuring of language.	Level I and Level III implications potentially in fundamental conflict as effective Level I military technology dramatically alters all elements of society in unpredictable ways (e.g., language undermined; privacy of thought eliminated)

But if I can build a helmet that knows what you are thinking, I can eventually build devices that can read your thoughts from a distance; and if I can do that, I can know your thoughts without your knowing that they are being read. From a Level II perspective, this capacity could offer important benefits, for example, airport security. Yet knowing what people are thinking is hardly tantamount to knowing what they are intending to do. If telepathic technologies were to replace, for example, the judgment of well-trained security personnel, would the result be better security? ⇐

As for Level III effects, consider what a telepathic helmet will communicate: not just nouns and verbs, but also images, and moving pictures, and fragments of sound, and perhaps olfactory and tactile memories and even emotional overtones and feelings. It is an extraordinary broadband communication system that we unlock here, and one that humans have never, in their entire history, had to contend with. What happens to language? Our communications systems and languages now are, in part, artifacts of the last jump in such technology: the printing press, which privileged the written word. English looks the way it does in large part because of that technology. What have e-mail and Twitter done to spelling and punctuation? What composite communication platforms have Facebook and other services already created? Even this relatively primitive beginning of social-networking technology is increasingly a world of multimedia transmission, with words and punctuation significantly modified by the requirements and capabilities of the technologies. What would a full-bore telepathic connection transmit? Language and communication, basic to the structure of any culture, would surely and unpredictably change.

And what about self? Although the boundaries may differ, we are all accustomed to three selves: a "public" self that the world sees, a "private" self that is limited to us and our friends, and a "no one else is ever, ever going to know what I'm thinking" self—the part of our mind that is never shared and that cannot be accessed with current technology. Would the capabilities of a telepathic helmet unlock this inner personal sanctum? The bandwidth is open. Proust wouldn't have to write volumes now; he could just transmit the sensation of the smell of the madeleine. A tongue-tied youth wouldn't have to struggle to find the right words to explain how much he loves; he just uploads and there he is. It would be like Facebook with an open broadband pipe directly into your self and your consciousness for all your friends—and, unfortunately, perhaps for everyone else. Consider the potential such systems would offer for hacking, for eavesdropping, even for rewiring other's brains through reverse transmission.

Remember that at present there is no way of knowing which of these possibilities will become real and which of them will turn out to have been laughable fantasies. But we have already started down the path. It is true that such radical transformations of the possible are nothing new; rather, they are an attribute of the techno-human condition. Yet experience also tells us that whatever does come of telepathic technologies will—because they are embedded in complex socio-technical systems, but also because of a normal human inability to sufficiently appreciate the contingency of cherished assumptions—be largely unanticipated and significantly unpredictable. Given that telepathic-helmet technology, like any other technology, carries its Level I, Level II, and Level III implications simultaneously, making decisions solely on the basis of Level I potential is bound to leave society unprepared to address, let alone prepare for, more complex unfolding consequences. And just because something sounds like science fiction doesn't mean scientists aren't working to make it non-fiction as soon as possible.

Case 3: Lethal Autonomous Robots

As the name indicates, our final example involves a suite of technologies with three relevant characteristics. "Lethal" means that the robot is equipped to kill human targets. "Autonomous" means that the decision to kill is the robot's, not a human's. (In other words, no human intermediary is in the loop of target identification, target verification, and target elimination.) What about "robots?" Most people agree that machines such as the Predator and Raven drones, the tracked platforms called Talon, PackBot, Warrior, and Swords used in Iraq and Afghanistan, and U.S. Navy's unmanned underwater vehicles are robots. But some people consider some land mines to be robots: they are inert until they sense the proper triggering conditions, then they jump, or explode, or do whatever else they are built to do. And what about a grid of surveillance/attack cyborg insects? Each cyborg insect, taken alone, may be too dumb to be considered a robot, but the grid as a whole may be intelligent. And what should we call a weapons platform that is wirelessly connected directly to a remote human brain? (In recent experiments at Duke University, a monkey with a chip implanted in its brain that wirelessly connected it to a robot in Japan kept the Japanese robot running by thought, so that it was in essence an extension of the monkey's physicality.) Even now, the Aegis computer fire-control system deployed on U.S. Navy vessels comes with four settings: "semiautomatic," where humans retain control over the firing decision; "automatic special," where humans set the priorities but Aegis determines how to carry the priorities out; "automatic," where humans are kept in the loop but the system works without their input; and "casualty," where the system does what it thinks is necessary to save the ship. Similarly, the Counter Rocket Artillery Mortar (CRAM) system is a computerized machine-gun system currently deployed to defend against rockets and missiles to which humans could not react in time.[9] Does the word "robot" signify a type of artifact, a type of capability, or a certain level of computational competence? Or does this discussion make the point that emerging technologies render dangerously contingent even words and concepts that we think we understand,

and thus leave us open to the risk of relying on implicit assumptions, discussions, and conceptual frameworks that are already obsolete?

Lethal autonomous robots (LARs) are controversial at every level, but discussion to date is often characterized by category confusion.

To begin with, why should such a technology, in any form, be deployed? The immediate response is Level I: "to save soldiers' lives." Indeed, robots, whether lethal and autonomous or simply robotic, do save the lives of soldiers: in Iraq and Afghanistan, many explosive devices that might otherwise have killed and maimed soldiers have been identified and eliminated by robots. But, as usual, the Level I answer only raises higher-order questions—after all, in World War I generals were willing to kill 100,000 men at a go by sending them into the teeth of concentrated machine-gun fire. So there seems to be some cultural contingency (influenced, of course, by the state of technological capability) at play here.

The U.S. military in particular faces a stark dilemma. It is charged by American society with being able to project force anywhere around the world, under virtually any conditions— that is what the military of a global superpower does. Yet in the United States the politics of military action increasingly demand that virtually no American soldiers' lives be lost in carrying out this mission. And technology is making it possible to decouple projections of force (to protect alleged national interests), from loss of life in battle (which can swiftly mobilize political opposition if the threat to U.S. interests is not widely viewed as dangerous or existential). That's why the Predator, a UAV that can be operated over Afghanistan or Pakistan while being controlled from Las Vegas, is so popular with the U.S. Congress.

And then there are simple demographics: the American population (as with the populations of other developed countries) is aging. There are fewer young people to fill boots on the ground. The net result is that it becomes necessary to design toward better military productivity, with productivity measured as mission accomplishment per soldier lost. This is one reason why the U.S. Defense Advanced Research Projects Agency is a major funder of research on how to keep soldiers in peak physical condition longer (which is dual-use research insofar as it also provides the scientific and technical basis for radical life-extension technologies). The substitution of military robots for people is an exact parallel to the substitution of capital for labor early in the Industrial Revolution. Robots, in other words, are another expression of the search for efficiency.

So yes, LARs will accomplish the Level I function of saving soldiers' lives. But they will also be counted on to fulfill a Level II, if not a Level III function: projecting power when cultural and demographic trends militate against casualties. And they will perpetuate the long-term trends of substituting capital for labor and increasing the productivity of the individual (which doesn't happen, for example, if you need dedicated human teams for each Predator).

Because LARs raise fundamental questions of human moral agency and accountability, even their Level I functionality is complex. For one thing, the laws of war[10] require that combatants adhere to two requirements: discrimination and proportionality. A combatant must be able to discriminate between enemy combatants and civilians, and is not allowed to target civilians. A combatant is also required to use only such force as is necessary for self-protection and to accomplish the mission (proportionality). Opponents of LARs argue that robots cannot possibly discriminate between civilians and combatants, or exercise discrimination, and are therefore not permissible under the existing laws of war. Opponents also argue that, by further distancing humans from the act of killing, LARs will encourage military adventurism while insulating both civilians and the military from the moral consequences and implications of actions said to be carried out in their interests. Notice that this confuses a Level I assumption (that LARs will be in fact be employed and will achieve lethal combat effectiveness) with both a Level II question (whether LARs are legal or can be designed to be legal under current laws of war) and a Level III consideration (that the ability to avoid human casualties through use of LARs might change social and cultural mores regarding the appropriateness of war). Table 7.3 illustrates some of the Level I, Level II, and Level III considerations that arise from autonomous-robot technology.

The premise of the opponents—that human soldiers behave better under actual combat situations than robots might—would be difficult to defend on the basis of the history of real soldiers in actual wars. Indeed, proponents of LARs argue that there is no reason to believe that robots cannot be programmed to behave better than soldiers. Yet the question of agency and accountability remains troublesome—and suggests, once again, that relations among Level I functionality and the complexity of Levels II and III is deeply problematic.

A robotic force, including LARs, would save lives on the side that wields them, but the potential for loss of life is perhaps one of the most powerful reasons why societies, especially democracies, tend to eschew war. Minimizing the loss of life, therefore, may increase the possibility that nations would choose war as a response to a situation that could otherwise be addressed by negotiation or with other policy tools. (The belief that conquering Iraq would be a short and relatively painless operation, due to American technological supremacy, may well have contributed to the dynamic that led the George W. Bush administration into war in 2003.) LARs also increase the psychological distance between those who decide to use force against other human beings and the results of that decision. Even today, operating a Predator and watching the flash that destroys a target is a very different psychological experience than shooting an enemy at close range, and perhaps even different than dropping

Table 7.3
Technology-level matrix for autonomous robots.

	Goals and effects	Implications for action
Level I: Military effectiveness	Legally project force without endangering humans (with or without humans in decision-to-kill loop)	Goals and technology align; therefore adopt technology. Save labor if LARS are deployed.
Level II: Adoption in civil society	Create technologies that can be substituted for humans in undesirable and low-paying jobs, especially in aging populations (e.g., Japan)	Implement technology, but technology alone may not lead to achievement of stated goal
Level III: Social and cultural effects	Substitution of robots for humans, and human-robot integration, create human varietals with greatly expanded capabilities (e.g., ability to operate in space environments, integrated human-machine cognition). Cultural implications not clear.	Level I and III implications potentially in fundamental conflict. Effective military technology, but effects of integrating the human with the robotic unpredictable at Level III.

a bomb from an aircraft. At the limit, killing an enemy becomes a video game (as in the science fiction classic *Ender's Game*); indeed, on the Internet today one can find a genre of videos called "war porn," in which images of real death and injury are juxtaposed with humorous dialogs or popular music. To what extent do the fundamental moral precepts that bind societies together become increasingly contingent as humans shed agency and accountability for acts of organized violence? ⟵

But these technologies may also lead in another Level III direction—toward the integration of the human and the robotic to create families of human varietals designed for different environments and behaviors. Remember that, in one of his most important insights, Adam Smith looked at a pin factory and realized that dividing the construction of a pin into different tasks, then having each worker specialize in a particular task, was much more efficient than having each worker make the entire pin. And the idea of division of labor was the conceptual basis for Henry Ford's assembly line, and for mass consumption more generally. So Smith's insight became a major factor in the Industrial Revolution and the rise of the modern world. Originally, of course, the concept applied to blue-collar workers. The rise of railroads and corporations on national scales—the trusts and the monopolies—in turn helped give rise to white-collar division of labor: the accountant, the human-resources professional, the corporate lawyer, the public-relations whiz. A combination of complexity, scale, and demand for efficiency drove these trends.

So what do we have now? In today's brutally competitive global economy, the homogeneity of the human species can be seen as an enormous inefficiency given the challenges and opportunities of the Anthropocene. Why depend on the luck of genetics, or the unpredictability of education and training, to match people to their tasks if you can design integrated human-technology systems that are optimally suited to a particular function? We humans, as wetware (biological organisms, made of protoplasm, that rapidly fail if the temperature gets too high or too low, or if we don't have food, water, and an atmosphere that suits us), are terribly designed for many of the places we might want to go (under the oceans, or to Mars, or even on battlefields). Human-robot combinations could be much more effective. So the society that successfully extends the principle of division of labor to the design of the human may well obtain an important comparative advantage that may simply be the next step in techno-human evolution. Perhaps some society will create integrated human/weapon platforms by coupling humans directly to robotic weapon platforms by means of wireless implants, or perhaps indirectly by means of plain old video controllers, or perhaps telepathic-helmet technology will simply be used to create a network for small-unit operations that integrates sensor technology with humans who function as components of an augmented cognitive structure. In any of these cases, the result will be a system that augments the plain vanilla wetware human in different ways, creating different capabilities. We do not, of course, say that such a path is desirable or ethical, but we think it would be foolish not to recognize that it is not radical.

Radical, perhaps. But in other ways the deliberate creation of human varietals is simply an incremental continuation of one of the most fundamental trends of the Industrial Revolution. For this latest step, one only need drop the assumption that there are non-negotiable barriers to changing the Cartesian, wetware human of the late twentieth century. And even if that step may be inconceivable in theory or in the ivory towers of academia, it is already happening in practice.

Concluding Thoughts

Our goal in this chapter has been to use some specific examples of emerging technologies to illustrate the complexity and challenge of understanding what technologies actually mean in their different domains—Level I functionality, Level II system complexity, and Level III Earth systems incomprehensibility. We chose military and security examples because much of the future is being forged today in military laboratories and tested on battlefields. The results of our brief survey are meant to be discomfiting: not only are the lines between science fiction and science fact pretty blurry; the lines between the different levels of technological complexity are blurry too, and there is no getting around this. The lack of ability to bring boundaries into focus is another attribute of the techno-human condition, and it cannot be addressed either by predicting beforehand (your predictions will be wrong) or by regulating afterward (either it will be too late or you'll end up regulating the easy stuff at Level I and missing the important stuff at Levels II and III).

We need to learn how to think, individually and collectively, much more effectively about technological complexity and humanness. The first step is to stop seeing technology as something outside our cultures and institutions, and to recognize it as a part of us. If nothing else, the transhumanism debates, and the rise of human-enhancement technologies, should make this recognition unavoidable, even if still deeply uncomfortable. But it also turns out that, once we have taken this step, a new path to understanding how to govern the consequences of our persistent and promiscuous inventiveness begins to open up in front of us.

8

In Front of Our Nose

Let's recap briefly. We started out hoping to make some sense of the debate over the virtues and pitfalls of technological enhancement of human capabilities, and quickly came to see that both sides were talking about worlds that no longer existed and probably never did—worlds of individual agency, of discernible cause-and-effect chains, of simple, unambiguous, fixed categories, of stable moral and metaphysical platforms. We explored the multiple levels on which technology makes itself felt in human affairs, and the contingency-generating, boundary-dissolving, yet often imperceptible ways in which human-technical systems continually reorder existence and are in turn reordered. After pursuing the meaning of the individual amid such complexity, we linked the continual emergence of apparently existential challenges at the interface of technology and society (climate change, economic meltdown, stupid wars, and so on) to failures to understand the essence of the techno-human condition—failures that characterize the debate over transhumanism, failures that motivated us to write this book in the first place.

Neurologists have a saying: "Diagnose, adios." So it often goes with books that try to look at the world in all its daunting incoherence. It's easy enough to analyze what's going on, but then what? One common trick is to devolve from reasoned critique to hortatory utopianism: "C'mon people now, smile on your brother, everybody get together, try to love one another right now."[1] Conversely, one can adopt the posture of absolute confidence: "If you're not with us, you're against us." In our line of work, however, the most common prescription is the further pursuit of Enlightenment rationality, achieved through some combination of two options: (1) "Do more research, reduce uncertainty, take action" and (2) "Educate the unwashed and the opposition." After all, if only everyone understood the facts, then the right course of action would become obvious—that is, either (a) embrace and promote technological change or (b) embrace and promote essential humanness.

We see these sorts of approaches (the utopian, the hard-ass, the rational techno-utopian, the rational techno-dystopian) as further symptoms of the world we have struggled to describe—a world unable (and perhaps increasingly unable) to come to grips with what it does to itself. We will take a different tack and offer up a modest set of attributes, for institutions (ranging from governments to research organizations) and for individuals, that we think would help constitute a world better able to manage the complex consequences of its own ingenuity. What makes our approach defensible is that we are not expecting fundamental changes in human nature, or redemption through technology. On the contrary. In our view, Dr. Pangloss was right. Well, half right. Well, a bit less than half right. The essential attributes of a society that can wisely address the ever-complexifying turbulence brought about by its own technological potency are right in front of our noses. The problem is that our Enlightenment instincts send us fleeing in the wrong direction, seeking knowledge and certainty (either scientific or metaphysical) when what is needed most is the courage and wisdom to embrace contradiction, celebrate ignorance, and muddle forward (but intelligently).

The idea that applied rationality can define a path toward solving problems that are increasingly embedded in complex and adaptive human-technological-natural systems created by

applied rationality is both the child and the assassin of Enlightenment ambition. As problems come to light (often as a
result of scientific inquiry), we continually seek to create more
systemic knowledge—more Level I technological control—
to help solve them, and we are continually suffused with a
sense of frustration about how much we seem to know, and
yet how little progress we seem to make. And then we complicate things by introducing Level II and Level III issues without
understanding that we are doing so, thus ensuring category
confusion.

Yet today our Enlightenment predilections give rise to ever
more shrill demands for rationality, quantification, factual
analyses, and evidence-based whatever, even as these demands
are a response to a world that increasingly defies understanding and management because of past commitments to these
same Level I analytical tools. We strive ever more assiduously
for control even as control more surely evades our reach. We
try to "Gosplan" climate change, biodiversity, and the ancient
cultures of the Middle East. Our problem is that we want to
turn everything into a problem that can be solved, when those
problems are in fact conditions. A cultural commitment to the
Enlightenment, to notions of rationality and inquiry and their
contribution to progress, encourages people to see deviations
from the world they would like to live in as invitations to make
the world better. The promise of rationality and inquiry, that
is, seduces us into interpreting various states of the world as
"problems" amenable to "solutions" that can be approached
through better understanding, more knowledge, more widgets.
We inhabit Level III, but we act as if we live on Level II, and we
work with Level I tools.

Is there an escape from this labyrinth? Trick question! (Alert
readers should be able to spot these by now.) Escape is not
an option, so let us reframe the challenge: What does it mean
to live successfully in this self-constructed labyrinth—to occupy all three technological levels simultaneously with some

elegance and agility, and, yes, with rationality, ethics, humility, and responsibility?

Here are some basic principles, derived from what we have said so far, for engaging the Level III world:

1. Eschew the quest for "solutions." Unpredictable future trajectories and challenges, dramatically increasing rates of data generation and learning, and ever-present divergences in values demand a continuous ability to adjust flexibly to new conditions. The more nimble our cultural, legal, regulatory, and social systems and institutions are, the better we will be at grappling with our evolving techno-human condition. What is needed is adaptability in the face of change, not stability in response to problems. This need has a personal corollary: as individuals, we should accept that we require some coherence in our psychology, our beliefs, and our worldviews, and that such coherence is bought at the cost of partial perception. For a long time we have sought truth as individuals, but now we know that were we actually to achieve truth we would no longer be human. That doesn't mean everything goes; it means that one must be agile, flexible, and able to recalculate when new data come in and when unexpected things happen.

2. Focus on option spaces. By this we mean that the unpredictability and complexity of the challenges we face are best met through a capability to adjust in real time, which in turn means we must have options available when our planned paths go off in wildly suboptimal directions. We need both technological options (what do you do when, for whatever reason, an important technology begins to cause more problems than it seems to solve?) and social options (how do we encourage institutions and social systems to think about alternatives, so that they can adapt quickly and agilely to new and unpredicted conditions?). Identifying multiple paths forward, and developing options before they are needed, will dramatically improve our ability to adapt. A relevant

example is provided by the electronics industry. When challenged to reduce use of stratospheric-ozone-depleting chlorofluorocarbons (CFCs), used for cleaning printed circuit boards and other electronic components, the industry did so fairly quickly. In contrast, when the industry was challenged to reduce the use of lead solder because of the toxicity of lead, it took many years to respond. The difference here is that in the case of CFCs a number of alternative technologies had already been explored and pilot tested but not used—terpenes, sophisticated detergents and other aqueous cleaning systems, and the like. In the case of solder, no real alternatives that met manufacturing and product performance requirements were available. In other words, in the first case there were options, even though no one had a clue about CFCs and ozone depletion; in the second instance there were no options. Of course, developing option spaces does not seem efficient for institutions that focus on maximizing short-term gain and analyzing problems in the Level I space. But if institutions favor muddling through over the long term above optimizing in the short term, increasing the availability of alternatives to important technologies and processes will be understood as enhancing their long-term prospects. The availability of option spaces, therefore, is at least somewhat responsive to cultural assumptions about efficiency and uncertainty.

3. Pluralism is smarter than expertise. Recall William Buckley's point about phone books versus the Harvard faculty. Governance should arise from many voices, rather than single authorities. This point is the social-system equivalent of option spaces: the more perspectives and voices contributing to social perception of, and responses to, unanticipated challenges, the more likely it will be that alternative paths can be developed and that robust and viable social responses can evolve. A single perspective, drawing forth a single representation of a complex underlying system, provides less

resilience in the face of unpredictable change than a number of different perspectives drawing forth different representations and different understandings, which then enable more rapid and rational adjustment in real time than would otherwise have been possible. Approach climate change from a purely environmental and environmental-science perspective and you don't convince the world; rather, you generate opposition from those whom you excluded from your dialog, but to whom you now presume to dictate policy. Quite incredibly, the private sector was just about completely excluded from early UN-based development of an international climate-change policy regime, despite the fact that the private sector is the dominant institutional player—with by far the greatest expertise about option spaces—in the production and generation of greenhouse gases.

4. Play with scenarios. This is another way to develop social options, and in fact literature and the arts, science fiction in particular, play an unrecognized and subtle role here. Not for nothing do institutions that live and die by adaptability, such as firms and armies, play games testing various assumptions and scenarios. They know that no real-world situation is likely to mirror their scenarios, but they also know that such activities provide experience in adjusting to unpredictable and rapidly changing situations

5. Lower the amplitude and increase the frequency of decision making. Many small decisions allow much more attention to be paid to complex systems as they evolve, so that policies can track the system more easily and more consistently and so that gaps between policy and reality don't grow dangerously large. Moreover, people and institutions tend not to get so invested in small decisions, whereas big decisions mobilize and lock in big constituencies. This sort of nimbleness is sometimes built into laws: in the United States, for example, relatively minor changes in regulations are rather easily accomplished, but for a major change a

cumbersome legal process of "notice and comment rule-making" is required. In practical terms, then, we should be creating a legal system for managing technological change that moves away from formal and procedurally complex processes and toward simple, transparent processes that produce interim results that can be changed as the context changes—and we should be getting stakeholders accustomed to such a legal system.

6. Always question predictions. Predictions are an extraordinary seductive type of information, especially when backed by sophisticated and incomprehensible science. They can make clear who will win and who will lose from a particular course of action, and they can shift responsibility from politicians (who have to make choices on the basis of value preferences) to experts (who can predict which decisions will favor which values). But we have explained why efforts to predict the future of Level III technologies are always wrong, and almost always wrong in ways that are surprising.[2] Questioning predictions is a way to ensure that the values, assumptions, and interests of those making the predictions, and those using them to support particular decisions, are brought out into the open. Institutions and societies that depend on predictions to make decisions about complex, evolving system conditions are introducing a source of rigidity and vulnerability into their deliberative processes that undermines the potential value of those processes. They are also turning over their decision process to an elite that, intentionally or unintentionally, can change policy and engage in social engineering by managing the data, models, and research programs that underlie such decisions.

7. Evaluate major shifts in technological systems before, rather than after, implementation of policies and initiatives designed to encourage them. This principle sounds straightforward, but people and economies tend to fall in love with particular technologies, and thus not to question their

potential for serious Level III consequences until the technologies are so deeply embedded in technological, economic,
and social systems that change is very difficult. Thus, for example, the United States and Europe moved to adopt biofuels on a large scale before any reasonable assessment was
done of the potential implications of such a policy. The U.S.,
which selected corn-based ethanol as the biofuel technology
of choice, seemed surprised when farmers shifted their crop
selection in response, prices for all food products rose around
the world, and food riots ensued in many countries. Remarkable. Another example arises from the dramatic increase in
reliance on Internet technology across major infrastructure
systems, which occurred in civilian systems because of the
increases in efficiency that better information management
provides. Unfortunately, reliance on Internet technology also
means that military opponents can compromise your infrastructure, without the need for a messy physical attack, simply by subverting your information structure with "back
door" technology. Had this vulnerability been recognized
earlier, more robust design might have reduced the potential for damage. But no one thought to ask.[3] In many cases,
of course, technological evolution is already occurring as a
result of powerful and unconscious cultural and economic
forces, but it is still possible to try to evaluate potential implications for the environment and for society so that the costs
can be minimized and the benefits maximized. For example,
the Internet, with its social networking, augmented reality,
nearly infinite memory, immediate accessibility, and information overload, is significantly changing human cognitive
patterns in new and unpredictable ways. The time to begin
studying these changes is now, as the technologies are being
developed, rather than later, when we may come to regret
some system-scale effects that nonetheless resist change because of technological lock-in, vested interests, development
of standards, network economics, and other phenomena.[4]

8. Ensure continual learning. In view of the unpredictability and complexity of the systems involved, continual learning at the personal and institutional level must be built into any governance process. We can never assume that we "know" these complex systems, because they evolve too rapidly; we have to continue to test our social, economic, cultural, and technological choices against what is out there. Some experience with this approach has already accumulated. In the case of technological systems and their management, for example, Robert Pool (1997) points out that "high reliability organizations" (such as aircraft carrier operations, air traffic control systems, or well-run nuclear power plants) have explicit continual learning structures to ensure that problems, errors, and opportunities for improvement are spotted and addressed. In response to complex natural-resource-management challenges in places such as the Baltic Sea, the Florida Everglades, and the North American Great Lakes, continual learning is embodied in an approach known as "adaptive management." A lack of continual learning in such efforts makes policy and social institutions quite brittle as environmental change occurs.[5] The learning process is messy, highly multidisciplinary, and often informal, but it is particularly critical in Level III situations. The process is tricky because the systems involved are complex, and because no individual, however well qualified, can understand them in their entirety—the learning goes on at the level of the institution rather than the individual. Indeed, it is the study of such mixed systems—human individuals and technologies acting as integrated cognitive networks in performing complex tasks such as navigation—that gives us some idea of how learning occurs in anthropogenic systems.[6]

9. Do not confuse economic efficiency with social efficiency. Economic efficiency can be measured, and Level I technologies often enhance it. If you install a machine that works better and uses less energy than its predecessor, you have

increased economic efficiency. But social efficiency is a Level III sort of beast: because it exhibits wicked complexity, it cannot be measured in quantitative terms. For example, one of us once did an in-depth study of solder substitutes as a test of early industrial ecology and "Design for Environment" processes. Many of the relevant economic efficiency metrics were obvious—product and process performance and cost, increasing or decreasing use of materials, and so forth. But some of the options would have stimulated significantly more mining than others. Socially, is mining good or bad? That question cannot be answered definitively, although there are ways of operating mines that support stronger communities, and ways that are destructive of human values and community (for example, the unregulated mining that goes on in many poor countries, often using the modern equivalent of slave labor). As in the case of expanded option spaces, social efficiency may arise from what appear to be economic inefficiencies: too much dialog, too many stakeholders, no central imposition of solutions, and the like. But the sort of advances that muddling produces, though themselves difficult to predict, may be socially efficient under most circumstances. And we think this principle is often well illustrated by successful business and military institutions. They may use economic efficiency or the military equivalents to evaluate some things—for example, whether to put a certain machine in a factory, whether to introduce a certain product line, whether to deploy a certain weapon platform. But to understand their social and cultural environment, they use scenarios and games and "serious play"—structured and intelligent muddling. Jane Jacobs makes a similar point in regard to the complexity of cities and economic development. Jacobs cites the observation of the Japanese anthropologist Tadao Umesao that "historically the Japanese have always done better when they drifted in an empirical, practical fashion . . . than when they attempted to operate by 'resolute purpose' and 'determined will,'" and she notes that Massachusetts' famous Route 128 technology corridor succeeded because of a "process of open-ended drift, taking up opportunities whatever they might be and whither they might lead."[7]

The Enlightenment program for human progress, especially as it is embodied in the quest for rational knowledge as a basis for action, encourages an approach to public discussion and action in the world that is largely an assault on every one of the precepts listed above. The Enlightenment approach glorifies the rigorous definition of problems, narrows options to arrive at a solution, moves decision making into the domain of experts, and of comprehensive action to "solve" problems. Consider, for example, the extent to which policy makers have handed off the climate-change issue to a relatively small group of non-governmental organizations, activists, and scientific experts, or the fight against terrorism to armies and the intelligence community. But "problems" and "solutions" are Level I concepts. When you inhabit Level III, you are experiencing "conditions," not "problems," and conditions are not states to be cured; they are, at best, to be accepted, understood, and wisely managed. And what gives us some hope that it may be possible to re-invent the Enlightenment—to move away from the concrete toward the inchoate, from the delusion of dominance and control to the thoughtful and reflective embrace of humility and tentativeness—is that in the real world, the Level III world, our suggested principles are reasonably strong descriptors of what actually happens. Complex problems drag on and on; many different groups and viewpoints advance many different, often competing solutions; action is forged from compromise and is rarely more than incremental, and then the whole painful process is repeated when conditions change so much that action cannot be avoided. The problem is that the Enlightenment lens views this sort of incremental muddling as a strong indicator of failure and primitiveness and strives to overcome it, thus pushing action in the wrong direction: toward waiting, doing more research, achieving more knowledge, relying more on experts, seeking more comprehensive solutions. And so we are constantly set back upon our heels as wave after wave of new technology breaks over us, and, like a drunk craving the hair of the dog, we prescribe more of what makes the problem worse. The most poignant exemplar of this addictive behavior today is the social response to climate change, in which enormous scientific, intellectual, political, diplomatic, and emotional resources have been mobilized to violate all our precepts on a magnificent scale in a top-down, knowledge-based control fantasy aimed at modifying the evolution of the coupled human-Earth system in particular ways. Those who have alternative framings are ignored if they can't be pilloried. Climate models segue from providing scenarios requiring serious contemplation to being treated as windows on the real future, while climate-change scientists become sources of absolute wisdom providing advice on restructuring social, cultural, and economic systems around the world. Many readers might disagree energetically with this portrayal, but we reiterate an empirically unavoidable embarrassment: after nearly 20 years of effort, this formalized, bureaucratic, and highly partial process has made *no progress* in reducing human emissions of greenhouse gases, even though the central article of faith and the main motivation in climate-change politics and policy is that such reductions are needed immediately. From the perspective of that framing, the problem only continues to get worse. Meanwhile, the identification and exploration of option spaces—particularly those that enhance the human capacity to be nimble in the face of a dynamic Earth system (in the parlance of climate-change, to adapt)—has been largely ignored. But that is a topic for another book (for example, Pielke 2010).

On a different landscape, military funding for human enhancement and surveillance technology continues to grow, driven by Level I considerations of effectiveness (especially in counterinsurgency environments) and by competition among states. But there is little if any corresponding attention being paid to potential Level III implications. Funding flows to specific technology development in the areas we mentioned in the previous chapter, and to many other similar technologies, but there is no funding available to ask the Level III questions, or to spin scenarios that might indicate that potential problems were, in fact, materializing, or to give us practice in thinking about how to manage such technologies over the longer term. Would it really be a reasonable tradeoff if we were to develop cyborg insects to help us to stabilize Afghanistan, but ten years later everyone here in the United States (and elsewhere) had lost any reasonable expectation of privacy anywhere a cyborg insect could go? If I could detect your very thoughts without your knowing it, who would benefit and who might suffer? How would economic and political power shift? We don't fault the U.S. military; their job is to protect Americans and to project power where civilian authority instructs them to, not to function as the technology assessor for the civilian implications of emerging technologies. But these are potent technologies not only in Level I military situations but also at Level III, where techno-human evolution is occurring right now. It offends reality to say that our collective responses to date are anywhere close to adequate. We emphasize, by the way, that this failure to engage with larger Level III implications is not just an issue with military organizations and personnel; we have heard ethicists approach any military technology as a tool of the devil, and lecture all and sundry about the evil of military activities, thereby ensuring that no one in the military will ever want to talk with them again—just as the corporate sector was left out of the early climate-change policy discussion. Thus each worldview seeks the coherence and perspective that only it can bring, in a world where all such worldviews are increasingly incomplete and inadequate. We return to Heidegger: "The flight into tradition, out of a combination of humility and presumption, can bring about nothing in itself other than self-deception and blindness in relation to the historical moment." Right now, we are all blind.

Forget about "solutions"; expand option spaces; expand the number of voices; make more frequent but smaller decisions; encourage questioning and continual learning, and dialog with Earth systems. This, we want to suggest, is the Level III way, but it asks for instincts totally contrary to those that emerge from Enlightenment commitments to applied rationality, individuality, and problem solving. Here Dr. Pangloss makes his return, for what makes these ideas more than just platitudes is that we do these things congenitally, but we tend to think of them as flaws—as conditions of our prelapsarian numbskulledness, to be escaped by means of the intellectual tools of the Enlightenment. So we are not proposing some idealized set of virtues, but a reframing of what we do anyway, making a virtue out of reality, and thus opening up the possibility of doing it more consciously, with more awareness of the choices we face and less tendency to lock in particular choices that seem logical now but turn out to look foolish down the road. Remember, it may be muddling through, but that is a skill and can be done badly or quite well indeed. Ask the Austrians at Königgrätz. And, as the final twist in the maze, we are also proposing that the one widely shared set of cultural precepts that is best suited to absorbing this reframed, humility-based approach to managing the techno-human condition is rooted in the Enlightenment itself. Only the spirit of the Enlightenment, with its commitment to not just truth but also pluralism and skepticism, could put up with the contradiction, ambiguity, and uncertainty that are the essential ingredients of the techno-human condition.

It wouldn't be the first time. Wrenching social change characterized the period during which the Enlightenment arose, and it has not been absent since—witness the rise of market capitalism and the subsequent Marxist critique thereof, the Darwinian reinvention of biology and of the relationships among living things, the Freudian reinvention of the self, the Nietzschean reinvention of philosophy, the Einsteinian reinvention of space-time. . . . In view of this effusion of foundation-shattering change, what are the necessary characteristics of a cultural system, such as the Enlightenment, that evolves successfully—and, more importantly, that is capable of functioning over a period of several hundred years that arguably has exhibited the most rapid economic, technological, social, and demographic change in human history?

The Enlightenment has succeeded as a global culture because it uniquely carries within it the seeds of its own negation as a uniquely "true" or "valid" culture. Despite the almost paranoid seeking for truth that characterizes Western culture, this search is understood to be always ongoing, and the resulting truths to be potentially expendable; in fact, the evolution of truth is central to the Enlightenment's own myths—Galileo undermining the truths of medieval Church doctrine, Kepler supplanting Copernicus, Einstein supplanting Newton, Bell supplanting Morse, Microsoft supplanting GM. The strongest critics of the Enlightenment have been its children—Rousseau, Marx, Freud, postmodernists of all stripes.

For these revolutionaries and critics, not only has the Enlightenment tradition been the source of the negation; it has itself been transformed, transcended, and made more universal and encompassing by the dialectic generated by the negation. Indeed, the Enlightenment framework succeeded—persisted— only to the extent it was able to continually negate itself as a unique source of "truth." But this process of self-negation was largely carried out in the domains of science and social theory, and largely in reaction to what had come before, not in anticipation of what might be coming.

Institutions and Anticipatory Self-Negation

What we want to suggest now is that the challenges of rapid and continual technological transformation require an acceleration of the life-giving process of self-negation that has allowed the Enlightenment, as a way of explaining and justifying certain types of human activity (especially the creation of knowledge and the accumulation of wealth), to flourish. We must move self-negation from a reactive, corrective role to an anticipatory role. With this in mind, we now add two precepts to our list of nine:

10. Intervene early and often. The best time to start talking about alternative technological pathways and perspectives is when ignorance is great and the horizon is fuzzy. As soon as technological capabilities begin to make themselves felt (usually through commercialization), vested interests (economic, political, emotional) begin to get organized, and people try to sniff out what the stakes are and whether they will be winners and losers. From that point on, open-minded discussion about decisions and policies becomes increasingly difficult to cultivate.

11. Accept and nourish productive conflict. Most ideologies either look back to a Golden Age or seek to create one—for example, Marxism, with its vision of the withering away of the State, or environmentalism, with its vision of sustainability. Yet humans are most adaptive, and most creative, in periods of bounded conflict—that is, when there isn't too much conflict (which brings chaos and destruction) or too little (which results in social stasis and a slow slide to irrelevance). Look at periods in which cultures bloom to create new spaces, new paths, and you always see contests: of ideas, of peoples, of interest groups.[8]

For many of the contested technologies relevant to our inquiry (vaccines, nuclear power, genetically modified foods, research on embryos, human enhancement), we can describe a standard evolutionary pattern, starting with early stimulation of vigorous debate, framed as a variety of pro-technology versus anti-technology ideologies that emerge from grounds ranging from raw economic self-interest to moral and religious precepts. Here we want to emphasize that early claims, both for and against a technology, are almost always rooted principally in ignorance about the future. Advocates of nuclear power promised energy "too cheap to meter"; opponents dreaded meltdown and proliferation. The reality, of course, has been much more complex and multi-faceted, but the most important point is that the technologies of nuclear power have evolved in response to both advocacy and opposition. Nuclear-power-generating technologies are much more diverse, safe, and reliable than they were 40 years ago, and one major reason for the improvement has been the dialectical relation between opponents and advocates of rapid deployment. (Such a dialectic was absent in the Soviet Union, and the result was Chernobyl.) Embryo research is telling a similar story. Advocates of embryonic stem cell research promise amazing benefits, at this point largely as a matter of faith, imagination, and the need to justify more research funding. Opponents, in their own creative space, see in the instrumental exploitation of embryos a cheapening of the value of life that will undermine civil society. One effect of these ignorance-based disputes has been to stimulate the search for alternative approaches to stem cell research that don't require destruction of the embryo. Option spaces are widening.

The organizational challenge is to take these sorts of unbounded and often pathological disputes and mainstream them—move them into the institutions and activities where technological change is created—into laboratories, universities, government offices, corporate boardrooms—*while ignorance is still rampant*. Our point is not to encourage ill-informed decisions and discussions, but to encourage, welcome, and embrace a capacity to reflect, at the early stages of technological decision making, on the choices that face scientists, technologists, and citizens, and, crucially, on why people make the choices they make in the face of profound ignorance.

Why should up-stream, ignorance-based reflection improve the capacity of people to grapple meaningfully with the techno-human condition? The answer is in part *procedural*: pluralistic, open, and aware deliberation is more democratically satisfactory than closed and clueless deliberation, and more satisfactory than a lack of conscious deliberation. If nothing else, it provides a sense of ownership and buy-in for subsequent decisions. And we are also raising the more intuitively challenging idea that the benefit is *instrumental*: that ignorance-based reflection moves technological change toward more socially desirable outcomes, and away from undesirable ones, as diverse decision makers reflect more deeply on the context of their decisions and on the uncertainty that pervades the context. And of course we fully recognize that this beneficial steering of technology can happen through a change in paths of technological change (nuclear reactors are safer and more reliable than they used to be), or through a change in the conceptions of desirability (climate change is forcing opponents of nuclear power to rethink their positions), or, more likely, through interaction of both kinds of change.

But we are also raising a third reason why ignorance-based reflection on technological complexity and change is valuable, and this reason is *emotional*: if culture encourages people to be more openly aware of the limits of knowledge, and of society's limited capacity to control the evolution of the complex socio-technical systems they create, then maybe, just maybe, people will also become less tolerant of stupid decisions that unnecessarily commit society to paths of action that ignore complexity. And perhaps they will become more tolerant of the seemingly inefficient and ineffective process of muddling through, of adopting rules and regulations and then adapting them as conditions shift.

One challenge is simply to make it safe to talk about technologies in terms of public values and choices, rather than in the simple input-output, more-is-always-better mode or in the equally simplistic more-is-always-worse mode (which demands that potentially useful Level I technologies always be banned because of hypothetical Level III implications). For example, it isn't hard to think of some fairly simple questions that could *always* be discussed when decisions are being made about emerging science and technology. In the face of our boundless ignorance, we can nevertheless ask and answer these questions:

- What are the values that motivate a particular investment in science or technology?
- Who holds those values?
- Who is most likely to benefit from the translation of the research results into social outcomes? Who is unlikely to benefit?
- What alternative approaches are available for pursuing such goals?
- Who might be more likely to benefit from choosing alternative approaches? Who might be less likely to benefit?
- Have alternative scenarios (or models) been explored? If so, what do they say about the preceding questions?

Serious attention to such questions has been remarkably absent from discussions about technological enhancement of humans, rooted as they are in Level I Enlightenment metaphysics of individuality, cause and effect, and problem and solution. Yet the questions are not hard to think about; they are as appropriate for government hearings or media inquiries as they are for academic scholarship. The problem is that the habit of asking them has not yet been formed.

Habits do change, however. Important norms of scientific practice, for example, have evolved greatly in the past several decades. Human-subjects research, the use and treatment of animals in research, environmentally safe practice, and gender diversity and ethnic diversity in the scientific community have all become mainstream concerns of policy makers and researchers, whereas in the recent past serious consideration of such issues was often labeled anti-scientific.

Moreover, these changes in norms have evolved in conjunction with changes in institutional structure. For example, concern about the ethical governance of human-subjects research in the United States led to nationwide institutional reform in the 1970s. Every publicly funded research project involving human subjects in the U.S. is now monitored by an institutional review board (IRB) that must approve the research before it can be conducted, and must ensure that ethical principles such as prior informed consent are enforced. The fact that there are thousands of such boards operating in the United States shows that comprehensive institutional change is a reasonable goal. IRBs are far from perfect in protecting the rights of research subjects, sometimes greatly reducing the efficiency of conducting research and other times acting as little more than an institutional rubber stamp. Nonetheless, they are an accepted element of a scientific infrastructure that respects and protects human dignity.

Just as the IRB process is an accepted part of all human-subjects research in the United States, anticipatory self-negation activities could be institutionalized within innovation-producing organizations by requiring ignorance-based deliberative activities for all major public programs and projects related to transformational technologies. This capacity building could be funded by a small tithe, perhaps 2 percent, on research and innovation expenditures. And although such a scenario may seem, right now, ridiculously ambitious, one could easily imagine a time, perhaps several decades in the future, when every major institution involved in creating, disseminating, and regulating technologies would be engaged in a continual process of reflecting on the values and choices that are implicated in its world-transforming work. At such a time, it will seem bizarre and unpleasant that in the first decades of the twenty-first century, with waves of nanotechnology, biotechnology, robotics,

ICT, and cognitive-enhancement technology about to break over society, unleashing a new era of techno-human evolution and disorientation, the institutions creating the waves were still locked into a mode of Enlightenment thinking that allowed them to believe that they knew what they were doing.

Macroethics and Ethical Uncertainty

Having suggested a framework for institutional behavior in the face of Level III complexity, what will guide individual behavior? The obvious answer is "ethical standards rooted in Enlightenment values (liberty, justice, equality, and so on)," but the challenge to thinking about individuals and ethics ought to be obvious by now: when it comes to technological systems, the connections between decisions and outcomes are so attenuated as to render any notion of ethical accountability meaningless. A cognitive scientist working to incrementally improve integration of neurons and electronic circuits is no more accountable for the future impacts of human-machine hybrids—cyborgs— than Isaac Newton, formulating his laws of gravitation, was responsible for the use of those laws in calculating the trajectories of artillery shells and bombs.

But the fact that much thinking about ethics is constrained by homage to simplistic Enlightenment values in much the same way as is thinking about technology or policy does not mean that there is no reasonable path forward. To attain some clarity here, we can posit three ethical categories, analogous to the technology levels we explored above. For simplicity, let's apply these categories to people who actually create technologies—engineers.

The shop-floor level of ethics for engineers is their everyday working ethics, consisting of two major components: the codes of ethics established by professional organizations, such as the Institute of Electrical and Electronics Engineers and the American Society of Civil Engineers, and, of course, the personal ethical structure that individuals bring with them (don't make up data, don't steal data, don't sexually harass your colleagues and subordinates, and so on). Professional ethics is utilitarian and rule-based, whereas personal ethical systems can reflect many different frameworks (utilitarian, rule-based, religious, and so on). This level of ethics is well understood and accepted, although of course complexities arise in different real-world situations.

The second level of ethics focuses on the subtler interplay between the engineer and the institutional context within which engineering is done. This context brings in explicit economic constraints and social considerations, and so raises the sort of network issues and conflicts we have seen before with Level II technology systems. For example, a demand to produce a cheap, simple, lightweight car can result in a design that is vulnerable to fires and explosions under certain crash conditions (as happened with the Ford Pinto in the early 1970s). Or engineers may be tasked by society to design nuclear weapons. Or engineering in a culture of corruption may lead to buildings in earthquake or hurricane zones that are unable to withstand foreseeable stresses. In each of these cases, the Level I goal is apparent (to design a car, or a building, or a weapons system), but the Level II goals interact with the artifact to produce undesirable and often unpredictable results. An engineer can design an almost completely safe car; but it will never be built, because economic, social, and systems constraints make it infeasible. Another engineer can, for moral reasons, choose not to design bombs; but someone else probably will do the job. Thus, there has begun to be a recognition of "social ethics," differing from professional ethics, that arise from the institutional context within which engineering is practiced.[9]

But just as the understanding of technology systems at the level of Earth systems raised issues far beyond those raised by simple networked systems, so too with ethics. Because the future paths of Level III technology systems are uncertain and unpredictable, ethical systems predicated on evaluation of future costs and benefits of a present action, such as utilitarianism, are unworkable—one simply can't identify or quantify the future stream of costs or benefits adequately. Moreover, as we have already discussed in the case of scientific models, Level III systems are complex enough so that any single ethical perspective can only be partial, which means that coherent rule-based ethical systems are also of limited value, because any particular rule-based system can provide only a partial perspective.

Oh, sure, there have been plenty of attempts to paper over this gap by applying existing ethical approaches to ethics at the level of Earth systems and Level III technology systems. (We will call ethics for Level III Earth systems "macroethics.") A well-known example among environmental and sustainability thinkers is the Precautionary Principle, stated in the UN's 1982 World Charter for Nature: "[W]here potential adverse effects are not fully understood, the activities should not proceed." Aldo Leopold provides another example related to sustainability: "A thing is right when it tends to preserve the integrity, stability and beauty of the biotic community. It is wrong otherwise."[10] Participants in the transhumanism debates have interpreted particular religious traditions or worldviews to require the banning of certain kinds of research and development, as if the connection from lab bench to moral spillover in society were knowable in advance. Any moral framework is incoherent if it seeks simply to extend existing ethical systems into more complex domains by, for example, making individual engineers or scientists personally responsible for the behavior of the larger technological systems on which they work. On the other hand, market fundamentalism, which posits (impossibly) that all economic transactions should be unencumbered by government interference, has become a Level III moral framework that acts in opposition to the above examples, as if unfettered permissiveness in the pursuit of technological innovation automatically leads to morally optimal outcomes.

Ethical frameworks that link Level I moral behavior to Level III knowledge assume not only that future technological paths can be predicted, but that single worldviews and belief systems are adequate to frame the ethical implications of complex adaptive Earth systems. Such rigid ethical frameworks are category mistakes. But there may be a deeper problem. An assumption underlying all these ethical formulations is that decisions about ethics by individuals or by specific political entities are meaningful because they will lead toward the desired consequences. Recent history suggests otherwise: neither the European Union's strong stance against agricultural genetically modified organisms nor the George W. Bush administration's efforts to limit federal funding of embryonic stem cell research has prevented the rapid advance of the science and technology in question—nor, for that matter, have anti-pirating laws prevented people from illegally downloading or copying software or music. Many technologies, including pharmaceuticals, supersonic air transport, and nuclear power generation, are limited by regulatory actions, but rarely has a society been able to forgo a powerful technological capability unless (as in the case of nuclear power, for example), there was an existing and economically feasible substitute.

Challenging all this may be, but the challenge is to the underlying assumption, not to the idea of an ethics appropriate for our Level III world. Ignorance is as pervasive in the ethical domain as in the factual domain. Who can know if the application of a certain set of ethical precepts to some complex domain of technology will actually end up supporting those precepts? For example, although we both suspect that, had we been of age during the early decades of the Cold War, we would have strongly supported efforts to prevent the development and proliferation of thermonuclear weapons, the historian Richard Rhodes (who wrote the authoritative histories of the development of the atomic and hydrogen bombs) has suggested, not implausibly, that the threat of mutually assured destruction helped to deliver 50 years of relative peace to the U.S., its allies, and the Soviet Bloc.[11]

Ethical uncertainty begins to look surprisingly like factual uncertainty when it comes to the techno-human condition. Neither ethical nor scientific analysis has much hope of predicting the future accurately enough to dictate appropriate behavior in the present. Moral dialog with a continually evolving and uncertain system means that different, even mutually exclusive worldviews are aspects of effective action.[12] But the path of effectiveness is not of the Enlightenment type (problem definition, factual certainty, moral clarity); rather, it is to be found in a recognition of the need for continuous dialog, for eternal vigilance—and, yes, for muddling as an important ethical process. We don't mean muddling as in relativism; we mean muddling as in understanding that ethics itself is an evolving system in a rapidly changing world. (It used to be the case, for example, that some people were not considered persons under some ethical and legal frameworks, a position that was strongly defended by people considered ethical at the time.) And we mean muddling as in understanding that ethics, like computer models and like worldviews, becomes partial as it becomes coherent— an ethical uncertainty principle that we cannot escape.[13]

A plurality of ethical and cultural perspectives reinforces the development of cultural option spaces and enhances not only ethical action in the short run but also the resilience of society in the long run. (Had the Scandinavian settlers of Greenland been able to shift to Inuit ethics and cultural patterns, perhaps they would have survived.) Similarly, if climate-change activists and scientists were more open to alternative worldviews and differing ethical prioritizations, they might not be generating the substantial backlash that is so evident today in the United States, in China, and across much of Europe. The challenge of al-Qaeda requires a difficult and ongoing balancing of security considerations, privacy, freedom of speech and religion, and a number of other moral and ethical domains; it is not advanced by claims that anyone who does not absolutely support a particular position is a traitor. But although differing and even competing moral perspectives are central to thinking about the role of macroethics in complex systems, this pluralism does not provide comfort to the individual-rights-based transhumanism argument that favors no-holds-barred human enhancement. That argument assumes that the utility of individual choice of enhancement will outweigh the costs at the individual, community, and cultural levels, a calculus that cannot be performed for a Level III system. But neither does it support the attempt to impose constraints by state action on enhancement technologies on the basis of religious or philosophical positions, which, given the public demand for enhancement, requires an appeal to proscriptive state intervention rather than to pluralistic political action. As with our previous discussion of institutions, macroethics dictates a process orientation to ethics, where the imperative is one of engagement itself.

That Level III macroethics shares the characteristics of unpredictability, uncertainty, and complexity with Level III technology systems does not mean that one simply gives up on ethics. Rather, it imposes different kinds of obligations. Thus, for example, individuals, governed by specific ethical codes and generally operating within particular cultural and ethical traditions, will continue to respond to Level I and Level II ethical challenges. But beyond that, they have an obligation to participate, and to encourage the appropriate institutions to participate, in evaluating ethical challenges arising from Level III technological evolution in the Anthropocene. Because macroethics requires ongoing dialog with systems that are changing unpredictably and in many dimensions (technological, social, natural, ethical, and economic, among others), individuals should support constant institutional engagement with such systems. Moreover, individuals bear an individual responsibility for contributing to ethical dialog, even though their perspectives are necessarily partial, because to fail to do so is to rob the system of its pluralistic wisdom.

The central weakness in the ethical framings of the transhumanism dialogs is now apparent: The idea of "ethical framings" is itself incoherent when the subject is Level III Earth system transformation, as transhumanism, however taken, certainly is. The assumption that ethical frameworks are stable is valid only in the short run. Accelerating technological and cultural change means that many of the questions and challenges that the anthropogenic world raises—including those associated with transhumanism—extend far enough into the future that the cultural models and assumptions on which ethical systems are built will themselves evolve and change. Humans will value (and recoil from) aspects of transhumanism technologies differently in the future than they do now. The specifics of those changes in ethical stance cannot be known. But, as our outline of macroethical considerations suggests, that does not mean we will be unable to conduct ourselves rationally, ethically, and responsibly under those circumstances—if we take the effort to learn how.

Individuality and Authenticity

And so, try as we might, there is no escaping the black hole pull of the monad, the individual consciousness, us. For ultimately it is we, as individuals, who must figure out how to thrive in the midst of the incomprehensibility we create. Given our enslavement to our own individual consciousnesses (at least until redesign), what lies between the Scylla of Level I narcissism and the Charybdis of Level III resignation and despair? What could it meant to engage authentically—fearlessly, openly, honestly—in a world that seems not only to render the individual meaningless but also to make comprehension impossible? It means that authenticity must build on a foundational cognitive dissonance. One must accept the validity of one's own experience, upbringing, culture, and other contributions to one's own grounding while simultaneously understanding that one is a partial and contingent reflection of the evolving and incomprehensible complexity that is out there.

The continual temptation offered by the Enlightenment is to escape this dilemma through resort to ideas and ideals of progress, especially the expansion of knowledge about our world. Yet, as we have emphasized, the cognitive networks we inhabit, and the systems we are seeking to understand, are not out there waiting to be revealed in ever more detail; they are created by the very queries we pose to the system, and the very cognitive network within which we gather data and process knowledge. Any framework or model that can be understood, and that is based on a coherent worldview, is by definition at best only a partial truth. Once could almost say "If you can understand it, it isn't True; and if it is True, you can't understand it."

Meaning, truth, and values, therefore, do not arise from first principles; they are functions of the state of the cognitive network—of our ordering of information and knowledge—and thus are contingent and continually regenerated in a reflexive dialog between cognitive systems posing queries to, and thus generating configurations of, external complex systems. Meaning, truth, and values are no more absolute than the network state that they arise from and reflect. But, we repeat again, that does not mean they are arbitrary. They are real—but only locally valid.

Each community, or ordered network, has values, rules, and behaviors that are valid within it, and also values, rules, and behaviors that would be destructive of it. The appropriate relativism is not "anything goes" but "anything (almost) may have its place." The challenge is to understand where the boundaries of your local system lie, and when its rules are valid and when they are not. The Scandinavian Greenland agricultural practices, for example, were coherent within the European community (and so useful and valid), but not the same as those of the Inuit. While the climate was stable, and allowed European style agriculture and Christian cultural practices, all was well.

But then the climate within which both societies functioned changed to favor the more flexible Inuit model. The Europeans could not understand the contingency of their own system of meanings, and they were doomed.

Meaning and truth, that is, arise from the dialectical process of their continued rejection. The challenge to individual authenticity lies in the quest for the integrity to create queries that are appropriate to the networks and contexts within which the individual is working. Just as the Enlightenment depends for its survival on anticipatory self-negation, so do the cognitive networks we create depend on the capacity of component individuals to engage in critical re-invention.

The guiding precept for individual authenticity must therefore be this: *That which you believe most deeply, you must distrust most strongly.*

To whom might we turn as an exemplar of such virtues of a re-invented Enlightenment? We offer as an exemplar George Orwell, who, notably among the intellectuals of the past century, strove to maintain both intellectual and moral clarity in the midst of chaos and contradiction. How did Orwell manage this? First, he was utterly unscientific—his observations and thinking were fraught with generalizations and assertions that were at best supported by telling anecdotes and were in any case completely untestable. ("People can foresee the future only when it coincides with their own wishes, and the most grossly obvious facts can be ignored when they are unwelcome. For example, right up to May of this year the more disaffected English intellectuals refused to believe that a Second Front [the U.S. invasion of France] would be opened. They went on refusing while, bang in front of their faces, the endless convoys of guns and landing-craft rumbled through London on their way to the coast. One could point to countless other instances of people hugging quite manifest delusions because the truth would be wounding to their pride."[14])

Second, he never shied away from making his own biases and preferences totally clear. The reader always sees the connections between Orwell's ruthless logic and his guiding principles. ("Any thinking Socialist will concede to the Catholic that when economic injustice has been righted, the fundamental problem of man's place in the universe will still remain. But what the Socialist does claim is that that problem cannot be dealt with while the average human being's preoccupations are necessarily economic."[15])

Third, he nevertheless (or, perhaps, therefore) often conveyed a powerful capacity to penetrate to the essence of what mattered, to see things as they were, achieving moments of crystalline insight by connecting aspects of the world that might have seemed unrelated. ("Snobbishness, like hypocrisy, is a check upon behaviour whose value from a social point of view has been underrated."[16])

And fourth, Orwell was relentlessly and painfully humble and self-critical. ("A man who gives a good account of himself is probably lying, since any life when viewed from the inside is simply a series of defeats."[17]) This explicit tying together of analytical and moral judgment ("I believe that it is possible to be more objective than most of us are, but that it involves a moral effort"[18]) is, of course, totally against the rules of the Enlightenment; but once we recognize that the world is as much constituted of moral (and other subjective) conditions as it is of the facts on the ground, this becomes a strength and not a fault. Indeed, the most important part of Orwell's analytical authority comes from his moral clarity (as contextual and contingent as that might be): we know where he stands, so we understand why he sees things as he does. Yet what makes this clarity most compelling is the *sotto voce* "of course I may well be wrong about this" that seems to shadow his every observation. The strength of his conviction and the depth of his humility are not a contradiction, but a synthesis that may look incoherent from an Enlightenment perspective that equates knowledge, control, power, certainty, and goodness, but emerges organically from a fearless confrontation with the techno-human condition.

Authenticity, we claim, requires that individuals accept and exercise their responsibility to critically engage the techno-human condition, while simultaneously accepting that their identity and even cognition are increasingly the products of emergent systems whose behavior cannot be traced through cause-and-effect chains back to the individual. We are all strangers to ourselves, not in Conrad's sense of always evading, even if unconsciously, the grim shadow of self-knowledge (as true as that may be), but in the sense of occupying worlds in which our own roles are beyond the reach of even the most advanced Enlightenment modes of inquiry.

The transhumanism dialogs are a mirror on a past that may never have existed but certainly will never exist again; they are, at heart, a vain effort to strangle the future with the dead hands of past verities. Within the boundaries of these dialogs, each individual is profoundly ignorant, and strives hard, above all, to remain ignorant of that ignorance. Having created our world, we now desperately pretend not to see it. But this is not a winning strategy: the complexity of the Level III anthropogenic Earth is not the far-off future, but is here, now. Sartre said "Man is condemned to be free." And, we would add, condemned to be a work in progress, a techno-human project in a constant state of reinvention, as we have been for thousands upon thousands of years. For now this freedom, from whence rises moral obligation, is neither comfortable nor (sometimes, at least) even bearable. But it is the freedom demanded by the historical moment, and it is non-delegable.

He, only, merits freedom and existence
Who wins them every day anew.

(Goethe, [1833] 1984, *Faust*, lines 11,575 and 11,576)

Epilogue: The Museum of Human Frailty

Recently we trooped off together to the Museum of Human Frailty. Housed in a restored factory building in a depressed mid-size Rust Belt city in upstate New York, the MHF's promotional brochure describes the museum's mission as helping "children of all ages understand their own emotional and rational contradictions and limitations."

After paying the modest entry fee, we entered a crowded exhibit room called the Hall of Memory. We were looking at maps of the brain projected onto the walls—standard science-museum fare—when someone yelled "pickpocket!" and a scruffy fellow ran for the exit. Chaos ensued. Afterwards, we sat in small groups with museum facilitators who asked us to recount events and identify culprits in a line-up. We gladly did so. Later, we watched a video of the actual event and saw how our memory sought to make order from the chaos, in the process getting most of the crucial details wrong and leading us to make false accusations based on poor assumptions and prejudices. We gained an appreciation of the processing power of our brains, but also of how little occurs at the conscious level.

In the Utopia Pavilion, visitors used computer simulations to try to solve real-world problems. This was our chance to save the world. But then . . . we got to see how our well-intentioned choices led to totally unexpected results, because it is not possible to anticipate and assess all the conditions of the situation at hand. For example, our decision to make the United States independent of foreign energy sources led to destabilization of several Middle Eastern governments, an explosion of regional conflict, a complete collapse of most major estuaries, and, 20 years out, a major shift in the nitrogen cycle that had huge impacts on biodiversity. But that wasn't what we were trying to do! Then one of us cured cancer and induced a collapse of the government health-care programs that President Obama worked so hard to implement, because of the high price of miracle drugs and the rapid increase in life spans. Of course, that dramatic increase in life spans completely wiped out our chances of energy independence—not to mention creating serious inequalities between those who were living to 150 and those who were still dying in their fifties. We were glad to see we were living to 150, until the fact that we hung around in positions of power long after we understood the world we were in, and blocked young people from being able to participate in any meaningful way in their communities, led our society to collapse. We hurriedly moved on.

The Gallery of Power included a virtual-reality version of a famous psychology experiment carried out at Stanford University in 1971. Museum visitors are put in charge of a room of holographic prisoners and asked by an authoritarian warden to keep them from misbehaving. Knowing little about the prisoners or the warden, visitors confront their own attitudes toward authority, and the limits of their own capacities to maintain moral integrity in the face of pressure to conform. Waterboarding is not an option, but we did put one particularly recalcitrant prisoner on bread and water, an act of weakness that now fills us with shame. We were encouraged in this action, however, by the enthusiasm of the warden.

By this time, our sunny optimism was turning to self-loathing, so we decided to try the Kondratieff Wave, a roller coaster whose course mimics the spasmodic ups and downs of economic markets over the past 200 years. Construction of the exhibit was completed in 2004, so the ride ends with a steep climb to the culmination of the housing boom (right after the brief but nerve-jangling dot-com-bubble free-fall). The MHF is now seeking funds to add another steep housing-bubble decline onto the end of the ride. The whole idea seemed a little gimmicky to us until one of the kids in our car gleefully remarked that, even though you can see the downhills coming, "it's still always a big surprise when it happens." After the ride, we watched a heart-wrenching 3D video about the rise and fall of the once-affluent city that is host to the MHF, now a cemetery of empty factories. But this bummer was followed by a video showing how new manufacturing centers were giving Asian citizens living in poverty a chance at their first reliable food supply, and even an occasional Internet connection, which was somewhat cheering—we needed it by that time. The curators obviously know what they are doing.

We entered the final hall—The Climatarium— with trepidation, anticipating a depressing experience. Sure enough, the exhibit began with projections of scientists and policy experts loudly explaining in great and painful detail how the planet was dying, a result unfortunately guaranteed by the failure of the United Nations climate negotiations, where the nations of the world sought desperately to agree on a binding treaty to follow the long-dead and much-mourned Kyoto Protocol. We could have done without the Greek Chorus, but this was, after all, the last, and only, hope of humanity, biodiversity, and Earth. But as we walked though the hall, heads down, we noticed other side-exhibits. Major oil companies displayed their natural gas and hydrogen technologies. A large enclosure held one of the first ambient atmosphere carbon dioxide capture "trees," using nanomaterial technology to grab carbon dioxide from the air around us; that carbon dioxide would ordinarily be liquefied and sequestered (but here was being fed into the soft drinks machine in the lobby). A case next to it held low-energy substitutes for lighting, ventilation, and other building functions. A viscous bubbling algae machine looked uninviting, but the diesel that spilled from its small processing unit went directly into the museum's HVAC unit. A mechanical politician explained, as we passed, how his constituency was adopting virtual office technology so they could avoid unnecessary vehicle travel during peak congestion periods. We watched a major scientific conference being held in a virtual-reality environment; the physicist running it was a pulsing purple fuzzball, but otherwise it looked as boring as usual. A buzzing exhibit that took up a whole room, an impressive model of the electric grid of North America, explained how smart grid technology was enabling smooth integration of many different sources of power, and adaptation to demand spikes caused by substitution of electricity for dispersed uses of fossil fuel. We were feeling a lot better, with the gentle refrain of "As ye muddle, so shall ye reap" caressing our ears as we headed for the exit. But on our way out, a guy with a pocket protector in his wrinkle-free white short-sleeve shirt handed us a brochure with some alarming data about carbon dioxide leakage from underground storage sites, and the Lost Species Meter ticked over yet another thousand species. We recognized our mistake. We had wanted a solution, a resolution, a happy ending to take back to our hotel room with us. What we got, actually, was all we could get: a condition requiring constant attention.

Sure, after few hours at the MHF you wind up feeling a bit tired of your own limitations. Yet in the end the MHF is not a downer. Because we found ourselves confronting our own inadequacies just as hundreds of others were going through the same experience, we left with a burgeoning sense of generosity toward the community of imperfect beings that constitutes humanity. Moreover, we understood that what we took to be inefficiency—our flailing around as we faced so many changes, and as our silver bullets either didn't solve everything or, worse yet, caused some other problem we hadn't heard of—was in fact efficiency in the face of a messy, complicated, unintelligible world. What a bracing antidote to the sleek and engaging science and technology museums, those staples of the modern American city, those paeans to the power and control fantasies at the heart of modernity.

One hardly needs science museums to be convinced of the potency and significance of human inventiveness. After all, if there is one thing that humans manage to do well it is to advance technologically—to the point that the history of our species is often described in terms of this advance, from the Stone Age to the Information Age. On the other hand, there are lots of things we seem to have persistent difficulty with. If the current economic, environmental, and geopolitical crises facing the United States and the rest of the world teach us anything, it is that at times the human capacities for technical ingenuity and control are no match for (and are all too often enablers of) human behavioral, organizational, and cognitive incompetence. If we don't understand, embrace, and even celebrate our apparent incompetence, it is hard to see how we will ever learn to manage our scientific and technological prowess. To match its science museum, every city needs its own museum of humility, of ignorance and uncertainty, of the techno-human condition, to help us better understand how to act wisely, prudently, and compassionately in the world. Meanwhile, we highly recommend a visit to the Museum of Human Frailty. It isn't a perfect place—but then again, how could it be? (It also isn't far from Cooperstown, so you can visit the Baseball Hall of Fame in the same weekend.)

Oh, and you might be interested to know that the museum is now installing a Hall of Existential Catastrophes, which apparently will include displays about real catastrophes that, for various reasons, we failed to stop (ignorance, politics, apathy and so on; the Darfur and Pol Pot exhibits are right next to the Black Death display), and others that were predicted to be Existential Catastrophes but turned out just to be nagging, often painful conditions of existence (finite resources and the population bomb). We understand there is a huge fight going on among the curators about whether or not they should feature terrorism for the inaugural exhibition. And as we pen these final words, we understand that several curators are now making a case for deep-sea oil blowouts, while colleagues insist that the Gulf of Mexico oil disaster of 2010 was a blessing in disguise because it will lead to adoption of new, non-fossil-fuel energy sources. We suspect they will compromise on terrorism, but we really don't care what they end up choosing, so long as they keep on arguing.

Notes

Chapter 1

1. "Fairly safe" (editorial), *The Economist*, August 2, 2008.

2. The question must be phrased carefully for legal reasons. Use of drugs such as Adderall, a methamphetamine, and Ritalin (methylphenidate), used to treat attention deficit hyperactivity disorder, or modafinil (e.g., Provigil), used to treat narcolepsy, for cognitive enhancement purposes—to strengthen concentration and maintain wakefulness over long periods, respectively—is "off-label" use. It is illegal for drug companies to advertise an "off-label" use, but it is legal for a physician to prescribe a drug for "off-label" use. It is, however, illegal to sell or trade in "off-label" drugs that have not been specifically prescribed for the individual.

3. This language was on the original website of the World Transhumanist Association, www.transhumanism.org. In 2008, that site was replaced by www.humanityplus.org.

4. The Great Chain of Being is a conceptual framework of the Universe perfected in the Christian medieval period in Europe. It envisions a structured hierarchy, with pure spirit and perfection (God) at the top, and pure matter and imperfection (rocks and other materials) at the bottom. In between, in order, come all things; angels are next to God, while plants are above matter, and beasts above plants. Humans are at the fulcrum point, for they are both spirit and matter. The Great Chain of Being was understood as representing an order willed by God, so changing it, as implied by transhumanism, is seen by some as constituting blasphemy.

5. There are many interesting books on the particularly American technological optimism and sense of progress—the "technological sublime," as David Nye terms it. Among those that the reader might find helpful in seeking to understand the role of technology in modern global culture are Nye's *America as Second Creation* and *American Technological Sublime* and the work that started the exploration of this corner of Americana: Leo Marx's *The Machine in the Garden*. (For full bibliographic listings of works mentioned or cited, see the bibliography at the back of the book.)

6. Bainbridge 2007.

7. The interested reader can explore these arguments in Moravec's *Mind Children* (1988) and in Kurzweil's *The Singularity Is Near* (2005).

8. Miller and Wilsdon 2006, pp. 14–15.

9. Garreau 2004.

10. See, e.g., Abrams 1971; Lasch 1991; Nicolson 1959; Nisbet 2003.

11. "Anthropogenic" means generated by humans. It does not mean completely constructed by humans: some Earth systems, such as the Internet, are; some, such as the carbon cycle perturbations underlying global climate change, aren't. What it does mean is that human impact, intentional or not, is now affecting global systems, be they natural, built, social, or cultural, at all scales, and thus that the world as we must now live in it is increasingly affected by the choices, activities, and general milling around of a single species: ours. The term "Anthropocene" makes the same point. (See "Welcome to the Anthropocene," *Nature* 424 (2003): 709.)

12. Heidegger 1977, pp. 49 and 136.

13. As Karl Marx wrote in his 1852 article "The Eighteenth Brumaire of Louis Napoleon," "Men make their own history, but they do not make it as they please; they do not make it under self-selected circumstances, but under circumstances existing already, given and transmitted from the past." In this, Marx apparently draws from Vico's earlier comment in *The New Science* (1725), "It is true that men have themselves made this world of nations, although not in full cognizance of the outcomes of their activities, for this world without doubt has issued from a mind often diverse, at times quite contrary, and always superior to the particular ends that men had proposed to themselves. . . ." So it is today: we create the Internet, for example, but, with its synthetic realities, metaverses, mash-ups, increasingly in-

telligent search engines, social websites, and the like, it would be foolish to pretend we understand what exactly it is that we have wrought.

14. See, e.g., Ellul 1967; Mumford 1928, 1970; Winner 1977.

Chapter 2

1. Bacon, quoted in Noble 1998 (pp. 50–51, 52).

2. Bacon, quoted in Mumford 1970 (p. 117).

3. Kurzweil 2005, p. 9.

4. Stock 2003, p. 3.

5. Bostrom, quoted in Garreau 2004 (p. 242).

6. Throughout history, many technologies have been developed under the pressure of military exigency, then have expanded to the civilian and commercial domains. Indeed, this may be taken in some ways as a dominant paradigm in engineering as a profession. More broadly, the emphasis on economic efficiency and competition between states for dominance means that institutions, rather than individuals, are likely to be the key shapers of transhumanist technologies.

7. We are aware that this dates us. Our students now "facebook" or "google" their way through our lectures. Occasionally it turns out they're googling to check on aspects of the lecture, which leads them in many instances to suggest alternative views or correct factual slip-ups. Most professors seem to address this issue by banning computers in their classrooms, which saves face but at the cost of denying the increasingly networked culture and cognitive practices of our digital native students—not to mention weaker lectures.

8. See Hirsch 1976.

9. An informal poll by the respected journal *Nature* indicated that 20 percent of its readers had used cognitive enhancers—specifically methylphenidate (Ritalin), modafinil (Provigil), or beta blockers—for non-medical reasons to stimulate their focus, memory, or concentration. See B. Maher, "Poll results: Look who's doping," *Nature* 452 (2008): 674–675.

Chapter 3

1. For good introductions to the subject of progress, see Lasch 1991 and Nisbet 1994.

2. See, e.g., Mokyr 1990.

3. In his 1943 paper "A Theory of Human Motivation," Maslow suggested a "hierarchy of needs" with five levels: physiological (e.g., eating, drinking), safety (e.g., personal security), love and belonging (e.g., family), esteem (e.g., self-esteem, self-confidence), and self-actualization (morality and creativity).

4. Mumford 1928, p. 283.

5. First-generation CBIs in the form of wearable headgear are now available for less than $300. They enable gamers to control avatars in synthetic realities through mental activity alone. They are very low resolution—they enable only limited functionality—but the proof of principle is there.

6. We are aware that the newspaper imagery here is yet another pathetic instance of our inability to keep rhetorical pace with technological change.

7. That said, we are well aware of the fragmentation of large social systems into communities with very different relationships to, and thus very different rates of adoption of, enhancement technology. Particularly avid adopters and diffusers of advanced technologies include the military, professional athletes, students (especially those in competitive programs), and gamers. As the science fiction writer William Gibson has observed, "[t]he future is already here; it's just unevenly distributed."

8. Mumford 1928, p. 60.

9. Because technology, especially at Levels II and III, is an integrated social, institutional, cultural, and economic phenomenon, all these elements co-evolve rather than directly creating one another. Put another way, causality at these levels is usually not simple but complex, reflexive, and contingent.

10. The other of us wishes he had had this experience.

11. Source: http://www.ntsb.gov.

12. Winner 1977, p. 228.

13. Less portentously, we note that in the classic song "Mustang Sally," first recorded in 1965 by Mack Rice but popularized in 1966 by Wilson Pickett, the purchase of a car by a man for his girlfriend enhances her freedom by enabling her to be "runnin' all over town," to his discomfort.

14. Vaccinating against infectious disease has an inherently democratizing aspect: the greater the number of the people who get vaccinated, the better off everyone is. The benefits of herd immunity are a strong stimulus for public policies that encourage widespread and equitable distribution of vaccines. This is a rare case in which individual enhancement is causally linked via the technology to a wider communitarian benefit.

15. See, e.g., R. Kyama and D. McNeil Jr., "Distribution of Nets Splits Malaria Fighters," *New York Times*, October 9, 2007.

16. Hill, Lines, and Rowland 2006.

17. Source: http://www.fightingmalaria.org.

18. Source: http://www.who.int.

19. Source: http://www.malariapolicycenter.org.

20. See discussion in Sarewitz and Nelson 2008.

21. Brown 1987. We thank Guillermo Foladori for bringing this article to our attention.

22. This example also raises the vexed questions of language and culture, languages of intelligibility versus languages of identity, and the extent to which learning different languages should be encouraged or even required. We do not choose to follow this line of inquiry. Readers interested in a relatively non-technical treatment are referred to Crystal 1997.

23. See, e.g., Porter 1999; Guwandi 2007.

24. See Lantz et al., in preparation; Gortmaker and Wise 1997.

Chapter 4

1. "Singularity" is a term used by some in the transhumanist debate to identify the hypothetical point at which artificial intelligence systems leap ahead of human systems, creating conditions that are essentially unknowable and unpredictable from this side of the phenomenon. Needless to say, it's highly controversial even as a concept. (See Kurzweil 2005.)

2. Disciplinary frameworks based on reductionist paradigms contribute to a view of "Earth systems" as merely chemical, physical, or biological—climate or oceanic circulation systems, for example. To the contrary, we believe that "Earth systems" cannot properly be understood without recognizing that, in the age of the anthropogenic

Earth, such systems inevitably contain elements of natural, built, and human systems, and must be perceived in those terms if we are to properly observe, understand, and participate in managing, the emergent characteristics of such systems. (See, e.g., Allenby 2005, 2007.) Thus, for example, to try to understand the current climate system without understanding technology as a social phenomenon, or the deep philosophic, cultural, and psychological patterns behind current human production and consumption patterns, is patently incoherent—indeed, the dismal failure of the current climate change negotiating process is strong evidence of this incoherence.

3. There are a number of good survey books for those who seek more information. Among them are McNeill 2000, Smil 1997, and, for an earlier perspective, Thomas 1956. On some of the policy implications of an anthropogenic Earth, see Allenby 2005, 2007.

4. In fact, scientists are surprisingly uncertain about how many species there actually are, which makes conclusions about absolute shifts in biodiversity somewhat imprecise. Even as we write this, the traditional estimate of over 30 million species, developed in 1982 by Terry Erwin of the Smithsonian Institution by extrapolating from beetle populations exclusive to one tree species in Panama, has been challenged by an estimate of 5.5 million, based on beetle species in Papua New Guinea. See, e.g., *New Scientist* 2010.

5. The magazine *Science* published the genetic sequence, and defended the decision in an editorial. See Sharp, 2005.

6. October 17, 2005, available at http://www.nytimes.com.

7. Future paths of complex systems are generally unpredictable, but that doesn't mean they are unbounded. For example, it is not possible to predict the weather in New Jersey for July 4, 2020, but it is possible to predict with fair probability that it won't be snowing. Moreover, unpredictability expands rapidly with time for most of these systems; a weather prediction for July 4, 2070 would be far less certain within significantly expanded boundaries. Who knows what we will be doing to the climate by then?

8. Thoreau, for example, was profoundly distressed by the railroad and its impact on nature, although he appears to have differentiated between trains and the sense of unnecessary urgency they created (which he disliked) and railroad tracks and beds (which in a short poem titled "What's the Railroad to Me?" he seemed to view as just another cart track, along which swallows played and blackberries grew).

9. This summary of the effects of railroad technology draws heavily from Freeman and Louca 2001, on Kondratieff waves and technology clusters, and from Schivelbusch 1977, on the social and cultural effects. The railroad technology system is also an important element in Rosenberg and Birdzell 1986 (a more general and very readable study of industrialization in the West).

10. The best and most detailed discussion of this is Cronon 1991.

11. Quoted on p. 37 of Schivelbusch 1977.

12. "Space-time compression" refers to the effect of technology in changing human perceptions of the spatial and temporal boundaries of their world. Though here we discuss the railroad, the automobile and the jet airplane are notable steps in continuing reduction of our sense of space and time. The rise of synthetic realities, and the capturing of human experience in cyberspace environments with no real mapping to any physical or temporal system, may be the ultimate step, turning space and time into manipulable dimensions of designed, non-physical systems.

13. Both of these quotations are from page 57 of Ney 1994. They remind us of the significant difference between premature utopian or dystopian visions of technologies and their eventual realities.

14. See, e.g., Boot 2006.

15. The resulting economic and social changes included the split between industrial and primary-producing economies (which played an important part in creating the divide between the developed and developing worlds) and a notable intercontinental convergence of commodity prices. See Findlay and O'Rourke 2007.

16. Among the useful books on this shift are Nye 1994, Nye 2003, Marx 1964, and Noble 1998. Additional useful writings in this fascinating area of cultural studies may be found in Nicolson 1959 and in Abrams 1971.

17. And often, as with war or other catastrophe, at huge human cost. See, for example, Polanyi's (1943) portrayal of the remaking of social fabric in industrialized England, or Dickens' portrayals of industrial London.

18. The reference is to the famous characterization of capitalism as "gales of creative destruction" in Schumpeter 1942.

19. McNeill 2000, pp. 193–194.

20. "Engineering and aging," *IEEE Spectrum* 41 (2004), no. 9: 10, 31–35. For much more on this controversial possibility, see de Grey

2004. Aubrey D. N. J. de Grey is a well-known and controversial advocate of what might be called the "radical human life extension" school. Whether life extension is possible, and if so when it will be available, and how long lives will eventually be, remain highly contentious within the relevant research communities. Interestingly, other science and policy communities, such as those associated with sustainability, are generally not attending to these possibilities, despite their obviously challenging implications.

Chapter 5

1. Similar dynamics arise in law, especially in arbitration and litigation, as one of us experienced in many years of legal practice; what is perceived by the Enlightenment rationalist as inefficient communication is, in fact, the complex process of muddling through complex legal, emotional, and factual tangles to workable solutions that arise from the fuzziness of the discussion, not from any failure to perceive the "rational" more quickly. In fact, that is one reason why the practice of law remains an art rather than a science. Perhaps the most important immediate implication of the effect of emerging technologies on law involves recent scientific advances demonstrating that in many cases actions may be unconsciously decided upon well before they become conscious, that adolescent brains may not process responsible choice as adult brains do, and that there may in some cases be genetic predispositions to violent or otherwise illegal action—scientific insights that undermine naive conceptualizations of free will. The implications of such findings for legal systems arise because legal responsibility, especially for criminal acts, generally is assumed to derive from freely chosen violation of social norms. The science and technology behind transhumanism may severely impact the principles upon which diplomacy, law, and indeed social interaction are based, but the transhumanist framing of the problem is 180 degrees out of sync with the real problems: the technologies will not tame the complexity of human interactions; rather, they get sucked into and become part of that complexity.

2. Roco and Bainbridge 2002, p. 6.

3. Rowling 2005, p. 197.

4. Hughes 2004.

5. Buckley on *Meet the Press* in 1965, as quoted on page 82 of Keyes 2006. Buckley went to Yale; in view of his intellectual consistency,

we are reasonably sure he would have held the same view about that university's faculty.

6. Perhaps virtual realities could, in the future, offer a platform for exploring such possibilities, but even in such a case it is important to recognize that we can change or emphasize different aspects of our personalities but we cannot change the fundamentals of the human condition.

7. On this important and perhaps underappreciated shift in Western intellectual history, see Abrams 1971 and Nicholson 1959.

8. McKibben 1989, p. 77.

9. Leon Kass, the thoughtful cultural conservative who chaired President George W. Bush's Council on Bioethics, has been the most articulate advocate of this position. See, e.g., Kass 1997.

Chapter 6

1. A system is simply a bounded set of interacting, interdependent parts linked together by exchanges of energy, matter, and/or information. Especially in systems such as those we have been talking about, the critical question is where to draw the boundaries: one usually wants to include strongly interacting parts, and exclude parts that interact only weakly with the included ones. A "simple" system can be thought of as one in which all the relevant outputs generated by a given input can be known (through, for example, cause-and effect analysis), whereas with a "complex" system future states generally aren't knowable, or at least aren't knowable in detail, because of non-linear interactions among parts of the system, complex feedback loops within the system, significant time and space lags, discontinuities, thresholds, and limits, and a tendency for the system, and its subgroups, to constantly adapt to changing internal and external conditions. See, e.g., Allenby, *The Theory and Practice of Sustainable Engineering* (in press).

2. For details of the Beer Game, see Senge 1990. Senge makes a very important point in summarizing his discussion (p. 40): "When there are problems, or performance fails to live up to what is intended, it is easy to find someone or something to blame. *But, more often than we realize, systems cause their own crises, not external forces or individuals' mistakes.*" The Cartesian mind—the transhumanist and his or her opponents—focuses on the individual; post-Enlightenment rationality must learn to focus on the system.

3. And we predict that they will continue to fail. This possibility is increasingly accepted even by those involved in the negotiations. Christiana Figueres, appointed Executive Secretary of the UN Framework Convention on Climate Change in July of 2010, was quoted in *Nature* as saying "I do not believe we will ever have a final agreement on climate change, certainly not in my lifetime." But so many people and institutions have committed their resources and credibility to the process that it is highly unlikely to slow, at least in the short run. This is a form of "cultural lock-in": institutional and personal involvement and psychological commitment is such that structures continue regardless of potential or changing context.

4. We recall a presentation by a senior climate scientist who, in beginning his lecture, asserted that climate change was a moral issue, then proceeded to explain how science demanded a particular approach, beginning with the adoption of the Kyoto Protocol. Both the speaker and his audience failed to question the premise that scientific expertise necessarily translated into moral or political authority. The point here is not that this climate scientist was "wrong" about the ethical implications of climate change, but that he so easily extended his unquestioned scientific expertise into political and ethical authority—a classic category mistake. Scientists and engineers often try to reposition wicked complexity into static and dynamic complexity, because that is their zone of familiarity and competence.

5. For example, see Prins and Rayner 2007; Sarewitz and Pielke 2008.

6. By this example we don't mean to suggest that the market is sacred, or that there is any such thing as a "pure" market in the real world, or that creation and coordination of economic activity should take precedence over distributional fairness (however one chooses to define it), or that all states, from American to China to Cuba, don't regulate their economies in many ways. We merely note the train wreck that ensues when applied reason is matched against wicked complexity.

7. See Conquest 2000.

8. See, for example, Simon 1990.

9. See Diamond 2005, especially pages 192–193 and 243–245. Diamond discusses a number of instances in which shifts in worldview could have preserved societies as their environment changed, but the community simply couldn't bring itself to change, and so suffered or even died out.

10. Although it would not be necessary if this were still an area of scientific inquiry, we believe the climate-change issue has gotten so

normative that we should make it clear that we both agree with the basic physical models indicating how climate dynamics are affected by human activities in some ways. We are less sure that climate change is the dominant existential crisis of our age, as it has been positioned to be by activists and some scientists, to the exclusion of other scenarios. (What about terrorists setting off a nuclear weapon in a major American city, and the Americans going rogue, for example?) Moreover, even when considering the potentially grave risks of climate change, we reject the telescoping of the social, political, technological, cultural, and economic complexity of climate change into the superficial language of carbon footprints. Doesn't nitrogen matter anymore? What about land use? Economic development? Starvation? Waterborne diseases? And how does one possibly connect carbon footprint at Level I to controlling the behavior of the climate at Level III?

11. See "Baby levy plan to offset carbon emissions," *Herald Sun*, December 10, 2007, available at http://www.heraldsun.com.

12. See Roberts 2007.

13. See Ravilious 2009 and the accompanying editorial (titled "Cute, fluffy and horribly greedy").

14. Johnsson-Latham 2007.

15. Reported in Ritter, "Dissenters demonized in climate debate," *Arizona Republic*, December 23, 2007.

16. Ellen Goodman, "No change in political climate," *Boston Globe,* February 9, 2007, available at http://www.boston.com.

17. See, for example, D. Sarewitz and S. Thernstrom, "Climate Change Scandal Undermines Myth of Pure Science," *Los Angeles Times*, December 16, 2009; J. Leake, "U.N. Wrongly Linked Global Warming to Natural Disasters," *Sunday Times* (London), January 24, 2010.

Chapter 7

1. See, e.g., Boot 2006; McNeill 1984; Keegan 1993.

2. Boot 2006, pp. 466–467.

3. Projections of military force often have undercut long-term security. Examples include the Soviet Union in Afghanistan, and Germany and Russia in World War I. See, e.g., Boot 2006; Keegan 1993.

4. Whether the American military response in the field to the challenges of counterinsurgency and police action has been more agile and

effective than the domestic social and political responses of the U.S. government at home is an interesting question. See, e.g., U.S. Army 2007, 2009.

5. Concern about privacy and surveillance has increased for many reasons, including the rebalancing of domestic security versus privacy interests that has occurred in the U.S., in the U. K., and in many European states as a result of increased terrorist activity, and improvements in common surveillance technologies such as continuously operating video cameras in public places. See, e.g., Michael and Michael 2010.

6. Source: BAE Wolfpack website (www.baesystems.com). C4I is military-speak for "command, control, computing, communication and intelligence."

7. See, e.g., Mitchell et al. 2008; Callaway 2009; "US Army invests in 'thought helmet' technology for voiceless communication," at www.physorg.com.

8. Augmented cognition ("augcog") involves a wide suite of technologies that sense battlefield conditions (in real or virtual space, as appropriate), prioritize opportunities and threats, and communicate as appropriate to the warfighter for response (or, in a mixed warfighter–autonomous robot environment, support the most effective system response). See, e.g., the website of the Augmented Cognition International Society (www.augmentedcognition.org), which identifies the main driver for such technologies: the limited bandwidth of human consciousness—"The primary challenge with these systems is accurately predicting/assessing, from the incoming sensor information, the correct state of the user and having the computer select an appropriate strategy to assist the user at that time." (http://www.augmentedcognition.org) Augcog also has significant Level I uses in civil society. Many car companies, for example, are designing augcog into their future automobiles in recognition of the fact that, as populations age, more of the cognitive load of driving must be shifted from the aging driver to the car and transport infrastructure. See, e.g., N. Fleming, "Look, no hands: Cars that drive better than you," at www.newscientist.com.

9. These examples are from Singer 2009.

10. There are three main branches of the laws of war: legal justification for engaging in war in the first place (*jus ad bellum*), legal conduct during war itself (*jus in bello*), and international agreements on military conduct and technology (such as the Geneva Conventions). In addition, domestic law may govern aspects of declaring and conducting war (e.g., the requirement in the U.S. Constitution that war

be declared by Congress—a provision that can be avoided by calling wars "police actions" or by seeking from Congress only emergency authority to use force). Moreover, in part because of the unprecedented changes in RMT, RNC, RCS, and RMOC discussed earlier, one may reasonably question whether the laws of war as currently constituted, which primarily reflect Western history and concepts of war, are still viable, either in whole or in part.

Chapter 8

1. This line is from the song "Get Together," written by Chet Powers (a.k.a. Dino Valenti) and best known in the Youngbloods' 1967 version.

2. For a fuller treatment of these issues, see Sarewitz et al. 2000.

3. Such failures are systemic. Even today, the educational process that creates the civil, environmental, mechanical, and industrial engineering graduates who are designing ICT functionality into these systems are seldom if ever introduced to concepts of information security.

4. Standards arise when technology systems must work with other technologies (in which case the standard governs interfaces between technologies), or when standards are necessary for a technology to link more widely. An example of the latter is rail gauges: interconnectivity of rail systems required standards (Shapiro and Varian 1999, who suggest this example, also illustrate its strategic use in noting that the Finns deliberately chose rail gauges different from the Soviet rail system to help prevent invasion). Network economics arise when the value of a technology is disproportionately enhanced by the expansion of the network of users of that technology—the telephone, e-mail, and social networking services are obvious examples.

5. See, for example, Gunderson et al. 1995; Berkes and Folke 1998. In the case of the Everglades, the National Research Council has a panel that observes, and comments on, the progress toward established goals in the re-engineering of that system. (See, for example, NRC 2008.)

6. Hutchins (1995) does an excellent job of discussing this very complicated issue.

7. Jacobs 1984, pp. 221, 230.

8. This is a difficult insight, given how much of our mutual efforts (outside of the domain of economic competition) are directed toward

reducing conflict, rather than encouraging productive conflict. How to encourage the latter may well turn out to be the most important cultural competence of the twenty-first century, but it is also necessary to manage destructive conflict. Frankly, we aren't sure we know the difference at this point. On the general subject, however, we strongly recommend Hall 1998. Hall makes the general point that culturally productive cities—the Athens of Socrates and Plato, early Rome, Renaissance Florence, Elizabethan London—were all environments characterized by high levels of conflict, intellectual and most often cultural as well.

9. A good example of this literature is Devon 2004.

10. Aldo Leopold, quoted in Sagoff 1988.

11. Rhodes 2003.

12. Among other things, this implies that scientific determinations of validity that depend on interdependent models with similar assumptions must be regarded as useful for generating scenarios, but should not be considered definitive—the reason being that any such model necessarily uses a coherent, and thus limited, ontology to achieve the simplification of reality that is the goal of any modeling activity.

13. See, e.g., Johnson 1993.

14. Orwell 1968, p. 297.

15. Ibid., p. 64.

16. Ibid., p. 224.

17. Ibid, 156.

18. Ibid., p. 298.

Bibliography

Abrams, M. H. 1971. *Natural Supernaturalism: Tradition and Revolution in Romantic Literature*. Norton.

Allenby, B. R. 2005. *Reconstructing Earth*. Island.

Allenby, B. R. 2007. Earth systems engineering and management: A manifesto. *Environmental Science & Technology* 41 (23): 7960–7966.

Allenby, B. R. In press. *The Theory and Practice of Sustainable Engineering*. Prentice-Hall.

Augmented Cognition International Society. www.augmentedcognition.org/applications.

Bacon, F. 1627. *New Atlantis*. Kessinger.

Bainbridge, W. 2007. Converging technologies and human destiny. *Journal of Medicine and Philosophy* 32 (3): 197–216.

Beattie, A. 2009. *False Economy: A Surprising Economic History of the World*. Riverhead Books.

Berkes, F., and C. Folke, eds. 1998. *Linking Social and Ecological Systems: Management Practices and Social Mechanisms for Building Resilience*. Cambridge University Press.

Boot, M. 2006. *War Made New*. Gotham Books.

Brand, S. 1968. *Whole Earth Catalog*. Portola Institute.

Brown, P. 1987. Microparasites and macroparasites. *Cultural Anthropology* 2 (1): 155–171.

Callaway, E. 2009. Brain scanners can tell you what you're thinking about. *New Scientist* 2732. Available at www.newscientist.com.

Clark, A. 2003. *Natural-Born Cyborgs*. Oxford University Press.

Clark, R. A., and R. K. Knake. 2010. *Cyberwar: The Next Threat to National Security and What To Do About It*. HarperCollins.

Conquest, R. 2000. *Reflections on a Ravaged Century*. Norton.

Cronon, W. 1991. *Nature's Metropolis: Chicago and the Great West*. Norton.

Crystal, D. 1997. *English as a Global Language*. Cambridge University Press.

de Gray, A. D. N. J., ed. 2004. *Strategies for Engineered Negligible Senescence*. New York Academy of Sciences.

Devon, R. 2004. Towards a social ethics of technology: A research prospect. *Techne* 8 (1): 99–115.

Diamond, J. 2005. *Collapse*. Viking.

Ellul, J. 1967. *The Technological Society*. Vintage Books.

Figueres, C. 2010. Sound bites. *Nature* 465: 850.

Findlay, R., and K. O'Rourke. 2007. *Power and Plenty: Trade, War, and the World Economy in the Second Millennium*. Princeton University Press.

Freeman, C., and F. Louca. 2001. *As Time Goes By: From the Industrial Revolutions to the Information Revolution*. Oxford University Press.

Fukuyama, F. 2003. *Our Posthuman Future: Consequences of the Biotechnology Revolution*. Picador.

Garreau, J. 2004. *Radical Evolution*. Doubleday.

Goethe, J. W. von. 1833 [1984]. *Faust, Parts I and II*. Princeton University Press.

Gortmaker, S., and P. Wise. 1997. The first injustice: Socioeconomic disparities, health services technology, and infant mortality. *Annual Review of Sociology* 23: 147–170.

Gunderson, L. H., C. S. Holling, and S. S. Light, eds. 1995. *Barriers and Bridges to the Renewal of Ecosystems and Institutions*. Columbia University Press.

Guwandi, A. 2007. *Better*. Picador.

Hall, P. 1998. *Cities in Civilization*. Weidenfeld & Nicolson.

Harris, J. 2007. *Enhancing Evolution: The Ethical Case for Making People Better*. Princeton University Press.

Heidegger, M. 1977. *The Question Concerning Technology and Other Essays.* Harper Torchbooks.

Hill, J., J. Lines, and M. Rowland. 2006. Insecticide-treated nets. *Advances in Parasitology* 61: 77–126.

Hirsch, F. 1976. *Social Limits to Growth.* iUniverse.

Hughes, J. 2004. *Citizen Cyborg.* Westview.

Hutchins, E. 1995. *Cognition in the Wild.* MIT Press.

Jacobs, J. 1984. *Cities and the Wealth of Nations.* Vintage Books.

Johnson, M. 1993. *Moral Imagination: Implications of Cognitive Science for Ethics.* University of Chicago Press.

Johnsson-Latham, G. 2007. A study in gender equality as a prerequisite for sustainable development: What we know about the extent to which women globally live in a more sustainable way than men, leave a smaller ecological footprint, and cause less climate damage. Report to the Environment Advisory Council, Sweden. Available at http: // www.genderandenvironment.org.

Kass, L. 1997. The wisdom of repugnance. *New Republic,* June 2: 17–26.

Keegan, J. 1993. *A History of Warfare.* Vintage Books.

Keyes, R. 2006. *The Quote Verifier: Who Said What, Where, and When.* St. Martin's Griffin.

Kramer, F. D., S. H. Starr, and L. K. Wentz. 2009. *Cyberpower and National Security.* National Defense University Press and Potomac Books.

Kurzweil, R. 2005. *The Singularity Is Near.* Viking.

Kurzweil, R., and W. Joy, 2005. Recipe for destruction. *New York Times,* October 17.

Lantz, P., C. Shultz, K. Sieffert, J. Lori, and S. Ransom. In preparation. The impact of expanded models of prenatal care on birth outcomes: A critical review of the literature.

Lasch, C. 1991. *The True and Only Heaven: Progress and Its Critics.* Norton.

Leake, J. U. N. 2010. Wrongly Linked Global Warming to Natural Disasters. *Sunday Times* (London), January.

Marx, K. 1852 [1994]. The Eighteenth Brumaire of Louis Napoleon. In *Karl Marx: Selected Writings.* Hackett.

Marx, L. 1964. *The Machine in the Garden: Technology and the Pastoral ideal in America.* Oxford University Press.

Maslow, A. H. 1943. A theory of human motivation. *Psychological Review* 50 (4): 370–396.

McKibben, B. 1989. *The End of Nature.* Random House.

McKibben, B. 2004. *Enough: Staying Human in an Engineered Age.* St. Martin's Griffin.

McNeill, J. R. 2000. *Something New Under the Sun.* Norton.

McNeill, W. H. 1984. *The Pursuit of Power.* University of Chicago Press.

Michael, M., and K. Michael. 2010. Special section on uberveillance. *IEEE Technology and Society* 29 (2): 9–39.

Miller, P., and J. Wilsdon, eds. 2006. *Better Humans?* Demos.

Mitchell, T. M., S. V. Shinkareva, A. Carlson, K. Chang, V. L. Malave, R. A. Mason, and M. A. Just. 2008. Predicting human brain activity associated with the meanings of nouns. *Science* 320: 1191–1195.

Mokyr, J. 1990. *The Lever of Riches.* Oxford University Press.

Moravec, H. 1988. *Mind Children: The Future of Robot and Human Intelligence.* Harvard University Press.

Mumford, L. 1928. *Technics and Civilization.* Harcourt, Brace.

Mumford, L. 1970. *The Pentagon of Power.* Harcourt, Brace.

Nicolson, M. H. 1959 [1997]. *Mountain Gloom and Mountain Glory: The Development of the Aesthetics of the Infinite.* Cornell University Press.

Nisbet, R. 1994. *History of the Idea of Progress.* Transaction.

Noble, D. F. 1998. *The Religion of Technology.* Knopf.

NRC (U.S. National Research Council). 2008. *Progress Toward Restoring the Everglades.* National Academy Press.

Nye, D. E. 1994. *American Technological Sublime.* MIT Press.

Nye, D. E. 2003. *America as Second Creation: Technology and Narratives of New Beginnings.* MIT Press.

Orwell, G. 1968. *As I Please, 1943–1945: The Collected Essays, Journalism and Letters, Volume 3.* Harcourt Brace Jovanovich.

Parker, G. 1996. *The Military Revolution: Military Innovation and the Rise of the West, 1500–1800.* Cambridge University Press.

Pielke, R. A., Jr. 2010. *The Climate Fix: What Scientists and Politicians Won't Tell You About Global Warming*. Basic Books.

Polanyi, K. 1943 [2001]. *The Great Transformation*. Beacon.

Pool, R. 1997. *Beyond Engineering: How Society Shapes Technology*. Oxford University Press.

Porter, R. 1999. *The Greatest Benefit to Mankind*. Norton.

Prins, G., and S. Rayner. 2007. Time to ditch Kyoto. *Nature* 449: 973–975.

Pumphrey, C., ed. 2008. *Global Climate Change: National Security Implications. Strategic Studies Institute*. U.S. Army War College.

Ravilious, K. 2009. How green is your pet? *New Scientist* 204 (2731): 46–47.

Rhodes, R. 2003. Technology and death. In *Living with the Genie: Essays on Technology and The Quest for Human Mastery*, ed. A. Lightman, D. Sarewitz, and C. Desser. Island.

Rittel, H., and M. Webber. 1973. Dilemmas in a general theory of planning. *Policy Sciences* 4: 155–169.

Roberts, I. 2007. Say no to global guzzling. *New Scientist* 194 (2610): 21.

Roco, M. C., and W. S. Bainbridge, eds. 2003. *Converging Technologies for Improving Human Performance*. Kluwer.

Rosenberg, N., and L. E. Birdzell, Jr. 1986. *How the West Grew Rich: The Economic Transformation of the Industrial World*. Basic Books.

Rousseau, J.-J. 1754 [1964] *Second Discourse. In The First and Second Discourses of Jean-Jacques Rousseau*. St. Martin's.

Rowling, J. K. 2005. *Harry Potter and the Half-Blood Prince*. Scholastic.

Royal Society. 2009. Geoengineering the Climate: Science, Governance, and Uncertainty. Policy Document 10-09.

Sagoff, M. 1988. *The Economy of the Earth*. Cambridge University Press.

Sandel, M. 2009. *The Case Against Perfection: Ethics in an Age of Genetic Engineering*. Harvard University Press.

Sarewitz, D., and R. Nelson. 2008. Progress in know-how: Its origins and limits. *Innovations* 3 (1): 101–117.

Sarewitz, D., and R. A. Pielke, Jr. 2008. The steps not yet taken. In *Controversies in Science and Technology*, volume 2, *From Climate to Chromosomes*, ed. D. Kleinman et al. Mary Ann Liebert.

Sarewitz, D., R. A. Pielke, Jr., and R. Byerly, Jr., eds. 2000. *Prediction: Science, Decision Making, and the Future of Nature*. Island.

Schivelbusch, W. 1977. *The Railway Journey: The Industrialization of Time and Space in the 19th Century*. University of California Press.

Schumpeter, J. A. 1942 [2008]. *Capitalism, Socialism, and Democracy*. Harper.

Senge, P. M. 1990. *The Fifth Discipline*. Doubleday.

Shapiro, C., and H. R. Varian. 1999. *Information Rules: A Strategic Guide to the Network Economy*. Harvard Business School Press.

Sharp, P. 2005. 1918 flu and responsible science. *Science* 310: 17.

Simon, H. A. 1990. *Reason in Human Affairs*. Stanford University Press.

Singer, P. W. 2009. *Wired for War: The Robotics Revolution and Conflict in the 21st Century*. Penguin.

Smil, V. 1997. *Cycles of Life: Civilization and the Biosphere*. Scientific American Library.

Stock, G. 2003. *Redesigning Humans*. Mariner Books.

Thomas, W. L., Jr., ed. 1956. *Man's Role in Changing the Face of the Earth*. University of Chicago Press.

UN (United Nations). 1982. World Charter for Nature. www.un.org/documents/ga/res/37/a37r007.htm.

U.S. Army. 2007. *The U.S. Army and Marine Corps Counterinsurgency Field Manual*. University of Chicago Press.

U.S. Army. 2009. *The U.S. Army Stability Operations Field Manual*. University of Michigan Press.

van der Leeuw, S. E. 2000. Making tools from stone and clay. In *Australian Archaeologists: Collected Papers in Honour of Jim Allen*, ed. A. Anderson and T. Murray. Academic Publishing.

Vico, G. 1725 [1999]. *The New Science*. Penguin.

Visvanthan, S. 2002. Progress and violence. In *Living with the Genie: Essays on Technology and the Quest for Human Mastery*, ed. A. Lightman, D. Sarewitz, and C. Desser. Island

Winner, L. 1977. *Autonomous Technology*. MIT Press.

Index

Prosthetic arms, 17
Proust, M., 149
Prussia, 75ff, 130
Prussian Railway Fund, 75

Radiation, 66, 67
Railroads, 3, 71ff
  and American exceptionalism,
    74–77
  and industrial capitalism, 73
  and principles of technological
    evolution, 79ff
  and Prussian military, 75
  and psychological dislocation,
    73, 74
  and Schlieffen Plan, 76
  and telegraph technology 72,
    73
  and time, 34, 71, 72
"Railroad singularity," 64
Raven drone, 150
Reagan, Ronald, 114
Reformation, 3, 118
Registry of Standard Biological
  Parts, 68
Reliability, of organizations, 167
Religion of Technology, The, 18
Reproduction, assisted, 57
"Responsibility to protect," 137
"Reverse adaptation," 45
Revolutions in civilian systems
  (RCS), 140ff
Revolutions in military opera-
  tions and culture (RMOC),
  140ff
Revolutions in military technol-
  ogies (RMT), 137ff
Revolutions in nature of conflict
  (RNC), 137ff
Rhodes, R., 182
Rittel, J., 109

Robotics, 8, 80, 178
Robots, military, 150
Roll Back Malaria campaign,
  48, 49
Roman Empire, 132
Rousseau, J.-J., 173
Route 128 (Massachusetts), 169
Russia, 139

Sandel, M., 19, 21
Sandia National Laboratory, 89
Sardinia, 53
SARS, 39
Sartre, J.-P., 189
Science, as belief system, 110,
  206
Schizophrenia, 16
Schlieffen Plan, 76
Second Coming, 78
Second Life, 81
Shakespeare, W., 16
"Shock and awe," 76, 127, 135
Shop-floor activities, 51, 63, 65
Silicon Valley, 135
Simon, H., 120
Singer, P., 141
Smallpox, 16, 31, 47, 70
Smith, A., 97
Soviet Union, 114
Space-time compression, 74
Spice Islands, 129
Stalinism, 31
Stem cells, 3
Steppe warriors, 129
Stirrups, 84
Stock, G., 19
Struldbruggs, 83
Swedish Ministry of Sustainable
  Development, 122
Synthetic biology, 68ff
Synthetic reality, 82